SpringerBriefs in Applied Sciences and Technology

PoliMI SpringerBriefs

Marta Dell'Ovo · Alessandra Oppio ·
Stefano Capolongo

Decision Support System
for the Location
of Healthcare Facilities

SitHealth Evaluation Tool

Marta Dell'Ovo ⓘ
ABC
Politecnico di Milano
Milan, Italy

Alessandra Oppio ⓘ
DASTU
Politecnico di Milano
Milan, Italy

Stefano Capolongo ⓘ
ABC
Politecnico di Milano
Milan, Italy

ISSN 2191-530X   ISSN 2191-5318   (electronic)
SpringerBriefs in Applied Sciences and Technology
ISSN 2282-2577   ISSN 2282-2585   (electronic)
PoliMI SpringerBriefs
ISBN 978-3-030-50172-3   ISBN 978-3-030-50173-0   (eBook)
https://doi.org/10.1007/978-3-030-50173-0

This Springer imprint is published by the registered company Springer Nature Switzerland AG
The registered company address is: Gewerbestrasse 11, 6330 Cham, Switzerland

Foreword

A structured decision process is a key factor to the success of any organization and for bringing benefits to the society, particularly in the context of healthcare. This book entitled *Decision Support System for the Location of Healthcare Facilities: SitHealth Evaluation Tool* contributes to this. The book brings considerations to the complexity in solving problems concerning the location of public services as hospitals, by employing an advanced multi-criteria decision analysis (MCDA) method integrated with geographic information system (GIS).

With GIS, the model proposed provides an integrated knowledge about territory and the explicit consideration of the spatial dimension of the decision problems. By using this approach, the distribution of healthcare facilities within cities is not neglected in healthcare policies.

With the advanced MCDA method, the site selection for urban facilities, especially for healthcare facilities, can be approached solving this crucial topic related to healthcare, handling the side effects produced and dealing with the multiple criteria involved. Therefore, it considers simultaneously with economic issues, other dimensions such as functional, locational, and environmental. Also, with this MCDA method, the vision of different categories of stakeholders can be embraced.

The authors represent an important group in the scientific community of operations research and have received in 2017 the award of the best conference paper by the EURO Working Group on Decision Support Systems, which is related to the work published in this book.

I first met Marta at the 2016 MCDM Summer School. Then, she visited my research group for a short time, when my first contact with her co-authors has been established. Since the first meetings, I could realize Marta's skills for advanced research and the authors motivation for developing unconventional work.

It is with great pleasure that I introduce a book that reports outstanding academic research results that supports the decision process related to issues of the utmost importance to society, particularly in the current days.

Adiel Teixeira de Almeida
Professor
Universidade Federal de Pernambuco
Recife, Brazil

Preface

Today, even more than before, decisions need to be strongly addressed. In the recent years, many infectious diseases have occurred all over the world. Differently from the previous infections, the recently emerged coronavirus disease 2019 (COVID-19) appears to be more aggressive in terms of number of cases, countries affected, and mortality rate.

Actually, COVID-19 cases are almost 3.7 million with a 33.3% hospitalization rate.

The geography of contagion has pointed out some reflection suggestions: Cities and their local and global supply chains and networks can amplify the pandemic risk and diffusion; spatial variables, being the big part of the problem, play a central role as determinants for structuring the response and improving preparedness; the WHO social model of health is on the edge between current threats and new advancements of built environment settings towards public health and safety; managing trade-offs among multiple, overlapping, and conflictual objectives, as in complex and uncertain decisions, requires robust methodologies and reliable data.

Despite the lack of comprehensive and homogeneous datasets across regions as well as the asymmetric governance and policy responses, many hypotheses have been defined about the relationship among territorial conditions and disease diffusion. Climate, air pollution, healthcare system governance and organization, physical hyper-connections, commuting travels, urban density, and social vulnerability are some of the potential drivers that stress the importance of territorial perspective as analytical and strategical framework. By observing the Italian response to the outbreak and especially the Lombardy Region, one of the most affected territories by the pandemic, the location of healthcare facilities network, compared to other regional and international contexts, has been able to cope with the health emergency, managing so far to limit its extent in the areas with the highest population density. Therefore, the location of hospitals can be considered as a precondition for the success of a new facility and an inadequate evaluation can strongly determine the capability to react to infections diffusion and to health diseases.

For understanding how environmental, social and economic features can promote or inhibit disease growth, spatial analysis is essential. This kind of analysis are not new. Before and after the 1792 Finke's medical geography approach and the John Snow's spatial visualization of the London cholera outbreak in 1854, many are the studies that overlaid the location of disease with factors that are considered as its drivers. In addition to the emerging influence of built environment and living conditions on health, maps by combining different layers of information suggest explanations and provide arguments for decision scenarios towards infection inhibiting and preparedness improving.

In the recent years, many geospatial models that combine geographic information systems (GIS) and multi-criteria decision analysis (MCDA) have been defined and applied in several decision fields. The use of GIS together with MCDA has revealed the potentials of spatial decision support tools in the context of complex decision-making with relevant territorial determinants. The current notion of health, being at the centre of several inverse and converse interactions among physical environment and socio-economic conditions as determining factors rather than being anymore the absence of disease, represents a complex decisions problem. In order to address choices towards a balanced trade-off between conflictual goals, multidimensional evaluation approaches are crucial. Furthermore, spatial multi-criteria decision support tools, by considering both geographic data and decision-makers' preferences as inputs, are helpful when the decision problem under investigation involves several stakeholders and government levels and has direct as well as indirect territorial triggers and effects.

Among ill-structured problems, decisions about siting healthcare facilities involves environmental, social, medical, functional, and economic factors, as well as different categories of stakeholders, thus requiring an aiding evaluation framework for authorities to make location choices and to communicate results to the community in a transparent way.

Given the above emerging evaluation issues, the volume proposes a spatial evaluation model for supporting public health policies at two different levels: (i) the location of new healthcare facilities, when a finite number of alternatives has not been yet defined, or to identify the most suitable one among a specific set (ex-ante evaluation) and (ii) to rate the suitability of an already chosen area and, in case of a critical score, to solve or minimize weaknesses and threats (in itinere and ex-post evaluation).

The combination of MCDA and GIS could address decisions by considering simultaneously multiple dimensions of the problem under investigation, and the maps could facilitate to interpret and communicate the results, by aggregating and communicating several layers of information and data, thus being a robust support of legitimization and transparency for public authorities and the many stakeholders involved.

Given these premises, the book is divided into five chapters. The first frames the decision problem by focusing on the location of healthcare facilities (Chap. 1). The second introduces a spatial multi-methodological approach as medium to support the development of spatial suitability analysis (Chap. 2). The third deals with the location of healthcare facilities (Chap. 3). The fourth described the SitHealth evaluation tool, a multidimensional tool for rating the suitability of locations to host hospitals (Chap. 4).

Finally, the book suggests how to support decision-makers within the policy cycle (Chap. 5).

Milan, Italy   Marta Dell'Ovo
                                                                                                      Alessandra Oppio
                                                                                                   Stefano Capolongo

Contents

Chapter 1
The Location Problem. Addressing Decisions About Healthcare Facilities

Abstract The location of healthcare facilities is commonly considered as an ill-structured decision problem since characterized by the coexistence of several and sometimes conflictual criteria, the presence of different stakolders with their own viewpoints and the necessity to find a trade-off among dimensions involved. Given these premises the contribution is aimed at proposing a detailed analysis of the State of the Art (SoA) of the location problem under investigation through a deep review of scientific papers, existing evaluation tools and the researches focused on solving the hospital's site selection. The main purpose of this chapter is to underline existing gaps on this decision problem in order to define an evaluation approach useful to structure a multidimensional framework. In fact, the analysis of the SoA has been developed in order to highlight how previous researches managed the decision problem by pointing out methodologies applied, software used and in particular criteria defined.

1.1 Introduction

Location problems are characterized by a deep complexity due to the coexistence of several and sometimes conflictual factors, interrelated subsystems and externalities. (Capolongo et al. 2020; Lenzi et al. 2020; Azzopardi Muscat et al. 2020). Their solution requires a multidimensional and integrated approach, based on the contribution of different fields of knowledge with respect to the many issues to be considered, both qualitative and quantitative, such as the site layout and characteristics, the stakeholders and the role they play into decision, the capital investment, etc. All of these aspects change according to the different kind of facilities to be located, having each facility its own peculiarity. In addition to the intrinsic characteristics to consider, a special attention should be given to the extrinsic ones, namely the factors influenced by the location, such as the quality of environment, the safety, the security, the accessibility, etc. (Kumar and Bansal 2016). Different steps have to be processed in order to structure a robust decision. Firstly, it is necessary to set the decision problem, then the main objectives and finally the alternatives that contribute to solve it. Moreover, by specifying the values to be satisfied, suggests aspects more relevant to deal with and could help in identifying the alternative most suitable to meet the values elicited (Keeney 1992; Keeney and Raiffa 1993; Hammond et al. 1999; Keeney 2004).

Within this context, the location of healthcare facilities can be considered as a typical ill-structured decision problem since it involves issues belonging to different fields of research and there are several and sometimes conflicting stakeholders, whose instances should be considered as crucial (Dell'Ovo et al. 2018a).

The objective of the contribution is to propose a detailed analysis of the State of the Art (SoA) aimed at investigating the scientific literature, the existing evaluation tools and the researches focused on solving the hospital's site selection decision problem. In particular this chapter aims at underlying existing gaps on this decision problem in order to define an evaluation approach useful to structure a multidimensional framework. After this introduction, the Sect. 1.2 is focused on the description of the main phases of a decision problem, Sect. 1.3 on facing the location problem considering both public facilities and healthcare facilities, while Sect. 1.4 proposes a deep analysis of the SoA and some preliminary conclusions are presented in Sect. 1.5.

1.2 Problem Statement

The current research is cross-disciplinary between two main different domains: healthcare facilities design and location and Multi-Criteria Decision Analysis (MCDA). As it will be better explained by the analysis of the SoA, it is widely acknowledged the lack of a robust methodology able to solve the decision problem concerning hospitals' location, but more in detail the lack of guidelines capable to support Decision Makers (DMs) on this specific decision problem (Kahraman et al. 2003a, b; Soltani and Morandi 2011). Since the issue under investigation will impact the society as a whole, as well as the natural and the built environment, it becomes important to define a decision support system that can address the site selection of complex constructions such as hospitals.

Many of the scholars who have investigated this topic belong to Operational Research (OR), defined by Ormerode and Ulrich (2013) as "an activity done by people for people", i.e. the final aim is to support people in their activities (Brans and Gallo 2007). In fact, its final purpose is to predict system behaviour and to improve or optimize system performance in the field of complex situation. Within this context the Decision-Making is defined as a multistep process able to guide DMs until the final choice/selection of the most suitable alternative among those available, according to a set of values or elicited preferences (Kersten and Szpakowicz 1994). Decision-Making follows precise steps to support future actions, such as understanding immediately why we care about that action and how it is possible to achieve it by providing different solutions and then validating the result according to robustness tests. Moreover, Decision-Making involves a certain number of stakeholders and it refers to a specific environmental context (Kersten and Szpakowicz 1994).

Already in 1986, Fredrickson, contributed in presenting an overview about how authors, at that time, structured strategic decision processes and in fact, he achieved the conclusion that there were similarities in the analysed approaches. The decision process's characteristics can be synthetized in six phases: (1) *Process initiation* concerns a deep understanding of the problem and who is involved; (2) *Role of goals* clarifies the purpose to achieve; (3) *Means/ends* aims to explain the relation between alternatives and goals; (4) *Explanation of strategic action* considers how alternatives can result in strategic actions able to solve the problem; (5) *Comprehensiveness in decision making* recognises possible limits or constraints; and (6) *Comprehensiveness in integrating decisions* evaluates the level of integration and acceptance of the individual decision.

Following the Simon's Model (1979), de Almeida et al. (2015) propose a model developed in twelve steps in order to solve a decision problem:

- Step 1—Characterizing the DM and Other Actors;
- Step 2—Identifying Objectives;
- Step 3—Establishing Criteria;
- Step 4—Establishing the Set of Actions and Problematic;
- Step 5—Identifying the State of Nature;
- Step 6—Preference Modeling;
- Step 7—Conducting an Intra-Criterion Evaluation;
- Step 8—Conducting an Inter-Criteria Evaluation;
- Step 9—Evaluating Alternatives;
- Step 10—Conducting a Sensitivity Analysis;
- Step 11—Drawing up Recommendations;
- Step 12—Implementing actions.

The first five steps belong to a "Preliminary Phase" where actors involved in the process and the main objectives to achieve are recognised. Once this fundamental information has been identified, it is possible to define an appropriate set of criteria with specification about their measurement scale (i.e. how to measure them). After the problem has been framed, the steps 4 and 5 are related to the generation of alternatives aimed at solving the decision problem.

The second phase "Preference Modeling and method choice" involves step 6, 7 and 8 and in detail the most suitable model to solve the problem is chosen, alternatives according to their performances are evaluated and weights are elicited.

The third and last phase, "Finalization", concerns the final evaluation and aggregation of data previously detected (weights and scores), the development of a sensitivity analysis able to validate the results obtained, a critical analysis of the process in order to define useful guidelines and then, once the DMs received the necessary information to take a decision, the process can literally start.

As it is possible to perceive from the cited examples, even if called with different name, the phases that characterize a decision problem are quite fixed and similar. First of all, it is necessary to understand the problem to face and the actors involved, after it is possible to define the objectives to achieve and to frame a set of criteria. Alternatives are then generated and their performances evaluated with the aim to solve the established problem. Once the weights have been elicited and aggregated with the scores according to the selected model (aggregation rule), the result is obtained.

The robustness of the process can be verified by performing a sensitivity analysis to better guide DMs in their final choice which can be considered as the final phase.

1.3   Facing the Location Problem

Once the problem and the process to be developed has been stated, it is possible to proceed by analysing in detail the decision problem under investigation within this research. In this section it will be presented the location problem starting from the general to the specific: first of all, the location problem will be investigated without clarifying the type of facility; once understood the key factors and the most adequate methodologies to be applied, it has been possible to move the attention to healthcare facilities and the factors that affect their location.

1.3.1   The Location of Services

In siting facilities there are several dimensions to consider (Capolongo 2016; Capasso et al. 2018), as it has been already described in the introduction. There is a wide literature about the location of facilities and the typologies are various. In order to give a brief example about how scholars have faced the problem and which kind of methodology has been applied, Table 1.1 shows a selected overview on different cases studies aimed at locating facilities. The literature review[1] has been supported by the Scopus Database which allowed to narrow the research by the use of keywords: "Location" and "Public Services" or "Public Facilities". A total of 591 papers have been found out. The first sort has been carried out according to the title, resulting in 164 papers, the second one according to the abstract and, from 53 selected, 12 have been chosen as most appropriate and representative of the context under analysis.

It is interesting to notice how most of the papers solves the problem by dividing it in several criteria in order to make it easily understandable, and that the 50% of the total involves the use of Geographic Information System (GIS) since the location issue considers spatial elements.

Kahraman et al. (2003a, b) have immediately clarified how choosing a facility location among a set of alternatives is a multicriteria decision-making problem, and since decisions concerning effects on the environment are more and more complex, it becomes necessary to involve experts in the process and in detail they suggested the use of group decision-making (social choice). Cheng and Li (2004) analysed the problem under the point of view of the investment: "Which one is the best location for investment?". This aspect should be a priority for developers. Cheng et al. (2007) have considered the complexity of a construction project across its life cycle and focused on how decision-making could be a support for developers.

---

[1] The literature review has been developed in May 2016.

Table 1.1 Selected literature about the location of facilities

Authors	Year	Context	Country	Facility	Method
Kahraman et al.	(2003a, b)	Group decision making	/	Best facility location alternative	Fuzzy multi-attribute group decision-making
Cheng and Li	(2004)	Management	China	Retail location selection	MCDM
Cheng et al.	(2007)	Management	China	Shopping mall	GIS
Neutens et al.	(2012)	Geography	Belgium	Accessibility to services	GIS-based multi-criteria analysis
Yang et al.	(2012)	Hospitality management	China	Hotels	Location model: mono-centric models
Lahtinen et al.	(2013)	Geography	Finland	Libraries	Travel mode and CO_2 emission
Gimpel et al.	(2015)	Marine policy	German	Aquaculture in offshore wind farm	GIS + MCE
Li et al.	(2015)	Automation in construction	USA	Underground utilities	Geospatial 3G system and 3D probabilistic uncertainty bands
Latinopoulos and Kechagia	(2015)	Renewable energy	Greece	Wind farm	Spatial multi-criteria decision analysis
Wang et al.	(2016)	Social sciences	China	Urban fringes	Questionnaire
Kumar and Bansal	(2016)	Architecture	India	Building in a hilly region	GIS-based methodology
Noorollahi et al.	(2016)	Renewable energy	Iran	Wind farm	Spatial multi-criteria decision analysis

Neutens et al. (2012) have underlined the importance of the accessibility and how transportation disadvantages could result in social exclusion. Yang et al. (2012) have explored the location of hotel and highlighted the importance of location strategies in order to be competitive on the market and obtain good performances. Lahtinen et al. (2013) stressed also the location under the aspect of the longer distance. According to their opinion, essential services should be equally spatially distributed and a wrong location could influence negatively the environment. For Gimpel et al. (2015) the use of GIS-based framework is fundamental to solve spatial issues even when different group of stakeholders are involved. Li et al. (2015), beyond the location issues, have highlighted the phase concerning preliminary analysis of the site chosen in order to facilitate workers and speed up the process. Latinopoulos and Kechagia (2015) applied a multidisciplinary approach, considering both economic and environmental aspects. Wang et al (2016) argued about the location under the social point of view since it affects the whole society, and people interested should be involved in the final decision. For Kumar and Bansal (2016) the best area to site a facility is where a building can be established with a minimum use of resources and is adequate also for future expansion. Moreover, according to their research, there should be a well defined priority among the selected criteria. Noorollahi et al. (2016), as others scholars, underlined how the location should be guided also by the public demand.

According to this brief overview it is clear how the location of facilities is a fundamental task for the whole society and could have both positive and negative impacts on the environment. Two clear considerations arise from the literature:

- the facility location could involve both public and private parties with different needs and expectations (Owen and Daskin 1998);
- the site selection is different from one project to another, then different projects are expected to entail different methods for location selection (Cheng and Li 2004).

1.3.2   The Location of Healthcare Facilities

Moving the attention from the location of facilities to the location of healthcare facilities, it is fundamental to have clear in mind in particular the last consideration drawn in Sect. 1.3.1 concerning the peculiarity of each project. Locating hospitals is a difficult task for all the reasons previously explained but the impact on the society could be crucial, in fact, it is a decision which has to consider the long-term consequences. In the following sections the problem is going to be investigated according to the literature, trying to explain why the location of healthcare facilities can be considered an Ill-Structured Problem (ISP), and understanding how the problem has been solved in the past.

According to Simon (1979) an ISP is usually defined as a "problem whose structure lacks definition in some respect". A problem can be defined as ISP when it is not a Well-Structured Problem (WSP). It means that the problem domain still has to be explored and there are no defined structures to follow. Compared to WSP, ISP lacks of the following characteristics:

- "a precise set of criteria is defined;
- it possible to locate the problem state at least in one problem space;
- changes can be represented in the problem space;
- new knowledges can be represented in the problem space;
- if the problem involves the real world then the problem space will reflect the laws that govern the external world;
- all information and conditions requested are available" (Simon 1979).

In agreement with the analysed literature, since the location of healthcare facilities does not respect the listed criteria, it belongs to the sphere of the ISP. Kahraman et al. (2003a, b) and Solatani and Marandi (2011) discussed about the multidimensional nature of the problem including both qualitative and quantitative criteria that cannot be addressed with conventional tools; Burkey et al. (2012) underlined the presence of many different stakeholders, Lee and Moon (2014) highlighted the key role of geographic location and Kumar and Bansal (2016) the important relation of the proposed building with its surroundings.

Considering this brief overview, it becomes much more clear the aim of the contribution, the interest on this topic and also in which field of research it can be placed.

1.4 Facing the Location Problem

In this section existing evaluation tools namely focused on analysing the energy efficiency of hospitals are presented and a deep literature review on the location of healthcare facilities have been carried out. This step will allow later on to frame a possible set of criteria according to the aim of the research and considering their frequency.

1.4.1 Literature Review

Many scholars, especially in the last 15 years, have investigated this topic by applying different methodologies to specific case studies in order to solve the location problem. Most of the papers have investigated the decision problem according to a multiple criteria perspective and with the support of MCDA. Moreover, some of them have proposed an integrated approach based on spatial analysis combining GIS and MCDA. The research has been carried out using the Scopus database with different combination of keywords consistent with the aim of the research[2]:

1. "Site selection" and "Hospital";
2. "SMCA" or "Spatial Multicriteria Analysis" and "Hospitals";
3. "Location" and "Healthcare facilities";
4. "MCA" or "Multicriteria Analysis" and "Hospital";
5. "GIS" or "Geographic Information System" and "Hospital".

A first sorting has been performed by selecting the fields of interest and a second one according to the abstract. The focal point was to understand which are the methodologies more frequently applied and to point out the main used criteria. The combination of the selected keywords took into consideration some assumptions detected by the previous analysis, such as the complexity of the decision problem and the involvement of multiple criteria, its spatial nature and the necessity to be supported by a robust methodology.

---

[2]The literature review has been developed in March 2016.

The results of the literature review have been widely presented by Dell'Ovo et al. (2016, 2017, 2018b) and Oppio et al. (2016).

Noon and Hankins (2001), supported by the use of GIS, have located a Neonatal Intensive Care Unit in West Tennessee by considering the patient density and the market share as decision criteria. For Daskin and Dean (2005) the accessibility was a key factor, since a wrong location of healthcare facilities can affect the increase of deaths and diseases. Murad (2005), in order to plan public general hospitals in Jeddah City (Saudi Arabia) have used both spatial and non-spatial data. Wu et al. (2007) have combined the Grey Relational Clustering (GRC) method and a hierarchical clustering analysis aimed at evaluating the level of concentration of medical resources in the Taipei area by analysing the following different criteria: year; bed number; hospital ownership—private or public—and distribution district. Vahidnia et al. (2009) have considered six key factors to locate a new hospital in Tehran (Iran): locations of five possible sites; distance from arterial streets; travel time area to access existing hospitals; contamination; land cost; population density, by combining GIS and Fuzzy Analytical Hierarchy Process (FAHP). Instead Soltani and Marandi (2011) have been supported by Fuzzy Analytical Network Process Systems (FANP) and GIS to locate an hospital in the Region 5 of Shiraz metropolitan area (Iran) and the following criteria have been defined: distance to arterials and major roads; distance to other medical centres; population density and parcel size. Burkey et al. (2012) and Faruque et al. (2012) have focused their investigation on the level of accessibility by considering determining factors, while Abdullahi et al. (2014) have defined the distance from main services and attractive points as crucial variables. The problem of expanding medical services has been investigated by Chiu and Tsai (2013) supported by the AHP and by considering the following criteria: demand of medical services; population demand and density; construction costs; availability of land, transportation, support to administrative departments and sectors involved in the process, existing hospitals and possible expansion plans. Moreover, it is important to not disregard the evaluation of potential constraints. Lee and Moon (2014) have developed a socio-demographic analysis on the city of Seoul (South Korea) defining as main criteria: number of hospitals; total population; population aged >65 years; number of businesses; number of workers; road area; number of subway entrances and residential area. Among the negative aspects which could affect the location of healthcare facilities, the traffic deserves to be mentioned. Beheshtifar and Alimoahmmadi (2015), conscious about this threat, have evaluated how to maximize the accessibility and the suitability of the selected sites by combining GIS with a multi-objective genetic algorithm. Zhang et al. (2015) have focused on intrinsic characteristics, by collecting data about the number of technical staff, the number of hospital beds and the number of large medical equipment. The satisfaction of hospitals' users is another variable that has to be considered as it has been done by Du and Sun (2015) and Kim et al. (2015), whose studies have focused on the evaluation of patients' needs.

The literature review underlines a convergence of criteria and a plurality of disciplinary perspectives and methodologies. The mostly considered deal with population density, accessibility and proximity to services since one of the common instance is to maximise benefits for communities.

Once analysed the selected literature, papers have been further reviewed by the support of tables with the purpose of deeply investigate which methodologies have been applied, in which decision context and in particular which criteria have been used. In Table 1.2 it is possible to appreciate how the analysis of the literature has been performed, by investigating the author, the year of the publication, the location of the case study and the type of building, the decision problem and context, the aim of the research, the techniques applied, the software that support the analysis and the criteria defined. Moreover, in Table 1.3 it has been developed a further analysis aimed at investigating the criteria's frequency among papers under investigation.

1.4.2 Existing Evaluation Tools

Nowadays sustainability in architecture is a very important topic and also a debated one in the daily life. New regulations and laws are increasingly trying to limit the impact of constructions' life cycle on the environment and to improve the buildings' performance (Bassi et al. 2019; Buffoli et al. 2012, 2015). At the same time, the relationship with the surrounding context is a crucial aspect, especially for healthcare facilities.

There are different tools in the international scenario aimed at assessing the intrinsic characteristics of buildings and hospitals with a special focus on their energy performance. The location is included as a criterion, but without the right importance as suggested by the literature previously analysed (Daskin and Dean 2005).

In order to start from the tools aimed at assessing hospitals and rating their sustainability (Brambilla and Capolongo 2019; Buffoli et al. 2015), it has been developed a research focused on grey literature. Tools resulted and analysed for the purpose of the research are going to be further described below.

The LEED (The Leadership in Energy and Environmental Design) Healthcare has been launched in 2011 in USA. It is divided into seven sections, and in the first one "Sustainable Sites" there are information about how to avoid the development of inappropriate sites, to reduce the environmental impact caused by the location of a building on a site and the description of different types of transport.

The BREEAM (Building Research Establishment Environmental Assessment Method) Healthcare (2010) is a protocol for environmental assessment and it has been developed in UK. It is divided into ten areas and those related to the site selection are: Transport, Land Use and Ecology.

In Italy there are no real tools to certify the sustainability of healthcare services, even if there is a set of guidelines, as the METAPROGETTO DM 12/12/00 which defines ten parameters which designers should consider while planning new hospitals.

Moreover, in Europe and all over the world, there are many other tools aimed at evaluating the energy and environmental sustainability of buildings. Even if these tools are not specific for healthcare facilities, they could be considered as further examples to get involved in a methodology and to understand how different criteria

Table 1.2 Analysis of the literature review

Code	Title	Author(s)	Year	Location	Type of building/service	Type of techniques	Software	Context	Decision problem	Aim
1	Creating a GIS application for health services at Jeddah city	Murad	2007	Jeddah, Saudi Arabia, (ASIA)	Private health facility	GIS	ArcGIS	Urban planning	Use of GIS for one private hospital in Jeddah city	Creation of a GIS application to cover three main health planning issues which are distribution of health demand, classification of hospital patients and the definition of hospital service area
2	A location-based comparison of health care services in four U.S. states with efficiency and equity	Burkey et al.	(2012)	North Carolina, South Carolina, Tennessee, and Virginia	Hospitals	2000 census	Microsoft MapPoint 2004	Socio-economic planning	The effects of the location of hospitals on the geographic accessibility of health care	Examination of the efficiency and equality in geographic accessibility provided by hospitals

(continued)

Table 1.2 (continued)

Code	Title	Author(s)	Year	Location	Type of building/service	Type of techniques	Software	Context	Decision problem	Aim
3	Hospital site selection using fuzzy AHP and its derivatives	Vahidnia et al.	2009	Teheran	Hospitals	FAHP + GIS	ESRI ArcGIS	Environmental management	Hospital site selection	Development of a multi-criteria decision analysis process that combines geographical information system (GIS) analysis with the fuzzy analytical hierarchy process (FAHP), and use of this process to determine the optimum site for a new hospital in the Tehran urban area

(continued)

Table 1.2 (continued)

Code	Title	Author(s)	Year	Location	Type of building/service	Type of techniques	Software	Context	Decision problem	Aim
4	Optimal selection of location for Taiwanese hospitals to ensure a competitive advantage by using the analytic hierarchy process and sensitivity analysis	Wu et al.	2007	Taiwan	Hospitals	AHP-based evaluation model. Delphi method		Environmental management	Hospital site selection to ensure a competitive advantage	This study adopts the modified Delphi method, the AHP and the sensitivity analysis to develop an evaluation method for selecting the optimal location of a regional hospital in Taiwan to determine its effectiveness
5	Hospital distribution in a metropolitan city: assessment by a geographical information system grid modelling approach	Lee and Moon	2014	Seoul, South Korea	Hospitals	GIS grid modelling approach	ArcGIS	Health	Information for decision-makers concerned with the distribution of hospitals and other Health care centres in a city	GIS is a useful tool for analysing and understanding location strategies, an approach that should be useful for decision-makers concerned with the distribution of hospitals and other health care centres in a city

(continued)

Table 1.2 (continued)

Code	Title	Author(s)	Year	Location	Type of building/service	Type of techniques	Software	Context	Decision problem	Aim
6	Spatial analysis of rural medical facilities using Huff model: a case study of Lankao County, Henan Province	Zhang et al.	2015	Lankao County in Henan Province, China	Hospitals	Huff Model + GIS	ArcGIS10.0	Planning	Hospitals' distribution	To assess the spatial distribution of rural medical services by using geographic information systems (GIS) and spatial accessibility indexes
7	Spatial data visualization in healthcare: supporting a facility location decision via GIS-based market analysis	Noon and Hankins	2001	West Tennessee	Hospitals	GIS-based Market Analysis		Management	Hospitals' location	Application of spatial data visualization to support the decisions of locating and sizing a proposed Neonatal Intensive Care Unit (NICU) within a system's network of rural hospitals

(continued)

Table 1.2 (continued)

Code	Title	Author(s)	Year	Location	Type of building/service	Type of techniques	Software	Context	Decision problem	Aim
8	Spatial modelling of site suitability assessment for hospitals using geographical information system-based multicriteria approach at Qazvin city, Iran	Abdullahi et al.	2014	Qazvin city, Iran	Hospitals	OLS and AHP + GIS	ArcGIS	Planning	Hospitals' location	One of the most important facilities in a community is the hospital. Due to the complexity of clinical buildings, in order to find suitable site for the construction of hospital, it is required to use sophisticated analysis with consideration of large numbers of critical issues such as technical, environmental, physical and social issues and many others

(continued)

Table 1.2 (continued)

Code	Title	Author(s)	Year	Location	Type of building/service	Type of techniques	Software	Context	Decision problem	Aim
9	Location allocation modeling for healthcare facility planning in Malaysia	Shariff et al.	2012	Malaysia	Healthcare facilities	Genetic Algorithm (GA)	CPLEX version 12.2	Computer and mathematical sciences	Locations for the healthcare facility	Solution approach based on genetic algorithm to examine the percentage of coverage of the existing facilities within the allowable distance specified/targeted by Malaysian government
10	Applying hierarchical grey relation clustering analysis to geographical information systems—a case study of the hospitals in Taipei City	Wu et al.	2012	Taipei City	Hospitals	Grey relational clustering (GRC) method a hierarchical clustering analysis	PAPAGO! SDK software (GIS)	Healthcare management	Attempts to combine GRC and hierarchical clustering analysis.	Understand the degree of concentration of medical resources in Taipei
11	Using GIS for planning public general hospitals at Jeddah City	Murad	2005	Jeddah City	Hospitals	GIS	ArcGIS/Arcinfo Version 8.	Planning	Ways in which GIS can be applied in health status	To evaluate the spatial distribution of public hospitals at Jeddah city using GIS

(continued)

Table 1.2 (continued)

Code	Title	Author(s)	Year	Location	Type of building/service	Type of techniques	Software	Context	Decision problem	Aim
12	Decomposing geographic accessibility into component parts: methods and an application to hospitals	Burkey	2012	North Carolina, South Carolina, Tennessee, and Virginia	Hospitals	p-median model + GIS		Economic planning	Geographic accessibility of facilities	To show the degree of accessibility using p-median and GIS techniques. An example is provided using hospital locations in four southern US states
13	Applying Analytic Hierarchy Process to select optimal expansion of hospital location The Case of a Regional Teaching Hospital in Yunlin	Chiu and Tsai	2013	Yunlin County, Taiwan	Hospitals	AHP		Healthcare management	Location of hospital's expansion	Finding the best location for hospital expansion or relocation
14	A multiobjective optimization approach for location-allocation of clinics	Beheshtifar and Alimoahmmadi	2015	Tehran, Iran	Clinics	NSGA-II, TOPIS and GIS	ArcGIS spatial analyst	Planning	Location-allocation problem	To establish new healthcare facilities, their optimal number and locations should be determined

(continued)

Table 1.2 (continued)

Code	Title	Author(s)	Year	Location	Type of building/service	Type of techniques	Software	Context	Decision problem	Aim
15	Location Of health care facilities	Daskin and Dean	2005		Healthcare facilities	Location set covering model, maximal covering model and P-median model		Management sciences	Health care and related location literature	To give the reader a feel for the models that have been proposed and the problems to which they have been applied
16	Spatial analysis to locate new clinics for diabetic kidney patients in the underserved communities in Alberta	Labib Imran Faruque, Bharati Ayyalasomayajula, Rick Pelletier, Scott Klarenbach, Brenda R. Hemmelgarn and Marcello Tonelli	2012	Alberta, Canada	Clinics for diabetic kidney patients	SatScan analysis	SatScan software, ArcGIS10.0 software	Health	Locations for new healthcare facilities in remote regions	To identify clusters of underserved patients with CKD and identify potential new facility locations for consideration by decision-makers

(continued)

Table 1.2 (continued)

Code	Title	Author(s)	Year	Location	Type of building/service	Type of techniques	Software	Context	Decision problem	Aim
17	Hospital Site Selection Using Two-Stage Fuzzy Multi-Criteria Decision Making Process	Soltani and Morandi	2011	Shiraz metropolitan area, Iran	Hospitals	FANP + GIS	Network analyst of ArcGIS version 9.3.	Urban planning	Hospital site selection	To report the process and results of a hospital site selection within the Region 5 of Shiraz metropolitan area, Iran using integrated fuzzy analytical network process systems with Geographic Information System (GIS)
18	Location planning problem of service centers for sustainable home healthcare: evidence from the empirical analysis of Shanghai	Du and Sun	2015	Shanghai, China	Home Healthcare	Location model, sensitivity analysis	IBM ILOGCPLEX Program 12.6	Urban planning	Location planning problem	To investigate the existing problem of home healthcare in Shanghai and to find the optimum location planning scheme under several realistic constraints

(continued)

Table 1.2 (continued)

Code	Title	Author(s)	Year	Location	Type of building/service	Type of techniques	Software	Context	Decision problem	Aim
19	Feasibility study on an evidence-based decision-support system for hospital site selection for an aging population	Kim et al.	2015	Dallas, Texas	Hospital for Aging Population	Evidence-based decision-support system (POP) with a Geographic Information System (GIS)	Google Earth Pro	Urban planning	Hospital site selection and planning	The feasibility of the evaluation framework in supporting hospital site selection for an aging population

Table 1.3 Check-list of criteria's frequency

Code	Health demand	Types of patients	Accessibility	Land cost	Contamination	Population density	Population distribution	Services	Proximity to sewerage system	Noisy areas	Existing hospital	Other
1	×	×	×									Equality of the location
2	×		×									Travel time to existing hospital
3			×	×	×	×						Capital, labour: demand for hospital personnel
4	×	×		×		×						
5			Road, number of subway			×					×	Population aged >65 years, Number of businesses, number of workers, residential areas
6					Traffic		×				Hospital distribution	Boundary, number of technical staff, number of hospital bed

(continued)

Table 1.3 (continued)

Code	Health demand	Types of patients	Accessibility	Land cost	Contamination	Population density	Population distribution	Services	Proximity to sewerage system	Noisy areas	Existing hospital	Other
7		Patient's density						×				Patient travel patterns
8			×		Distance to polluted areas	×				×	Distance to noisy areas	Distance to existing hospital	Proximity to more populated areas especially for adults older than 65 years, size, distance to rivers and canals
9												Distance to existing hospital	
10			×										
11	×		×			×	×					×	
12			×				×					Distance to existing hospital	
13	×		Public, parking, main roads	×			×					Competition with existing hospital	Hospital construction, future development

(continued)

Table 1.3 (continued)

Code	Health demand	Types of patients	Accessibility	Land cost	Contamination	Population density	Population distribution	Services	Proximity to sewerage system	Noisy areas	Existing hospital	Other
14	×		Travel cost		Ground condition		×					Land use
15			×								×	
16		×	×				×					
17			×			×					Distance	Land size, traffic congestion, land shape, land use
18	×		Transportation time								×	Total investment cost
19	×	×	×				×	×			Healthcare services provided	Income, smoking people, frequency of visitation healthcare, water and electricity prevision, funds from government

are evaluated. According to the objective of the research, it has been moreover investigated the PROTOCOLLO ITACA (Istituto per l'innovazione e Trasparenza degli Appalti e la Compatibilità Ambientale) (2011), an Italian tool that is usually applied to residential buildings, offices, shopping centres and industrial buildings and the CASBEE (Comprehensive Assessment System for Built Environment Efficiency) (2006), a Japanese evaluation system based on the assessment of features such as indoor comfort, aesthetic, use of materials and energy consumptions.

Table 1.4 shows a comparative analysis where the main important extrinsic characteristics for assessing the location suitability for hospitals siting have been elicited and compared among existing tools.

Table 1.4 shows the frequency of criteria, that considers extrinsic characteristics of the tools previously analysed. According to the recurrence, it is possible to conclude how the more frequent and thus essential criteria are "Centre of urban redevelopment" and "Built-up areas", that encourage the use of built-up areas in order to promote urban regeneration process, and "Connection to green areas" that underlines the benefits of choosing sites located close to the natural environment rather than in highly dense areas. "Alternative transport" ranks second, since it promotes the use of soft mobility (e.g. bicycle) and discourages the private transport being hospitals public facilities should be easily accessible. Moreover, "Flexibility", interpreted as the possibility to change and expand internal spaces of the hospital, "Public transport", to discourages the private mobility and "Proximity to services", to integrate healthcare facilities with the supply of urban public service system and avoid isolation.

Even if the frequency has been taken into account, also the other criteria (most of them) can be judged fundamental in selecting a site for the location of hospitals given the results obtained by the analysis of the literature.

1.5 Conclusions

This first chapter proposes an overview of the main concepts at the base of the research and also an explanation of the problem statement.

Starting from the general definition of the location problem and in detail of the healthcare facilities and moving to the analysis of the literature review and existing evaluation tools, current strengths and weaknesses came to light.

With respect to the first reflections, it is urgent to remark how this research is trying to bridge the gap between two research and practice domains nowadays distant but with many contact opportunities: the OR and the health system.

Once understood the main topics, the location problem has been investigated supported by a literature review with the aim of pointing out the main differences among the several approaches and methodologies applied. One important issue resulted from this part concerns how the site selection is different from one case to another, then different decisions are expected to entail as much different methods for location selection (Cheng and Li 2004). Moving to the literature about the site

Table 1.4 Evaluation tools: comparative analysis

	Centre of urban redevelopment	Flexibility	Built-up areas	Building density	Public transport	Alternative transport	Car parking	Diversification of accesses	Proximity to services	Connection to green areas	Noise pollution	Air pollution	Unhealthy industries
1. LEED healthcare	×	×	×	×	×	×	×		×	×			
2. BREEAM healthcare	×	×	×	×	×	×	×		×	×	×	×	
3. Metaprogetto	×	×	×				×	×		×			
4. Protocollo Itaca	×		×		×	×				×			
5. CASBEE	×		×			×	×		×	×	×	×	×
Frequency	5	3	5	2	3	4	4	1	3	5	2	2	1

selection of healthcare facilities, criteria and methodologies used in each paper have been identified and some important features detected. To solve the problem, most of the scholars have used MCDA and GIS, sometimes combined, given the multiple aspects involved and their spatial nature.

The investigation of existing evaluation tools has allowed both to identify criteria and to understand how they have been measured.

From this first problem investigation, some criticalities have emerged as the lack of a specific location trend and lack of a defined set of criteria to use, whose determination is in fact connected to specific requirements and conditions of the context. On the contrary, for what concerns the methodology, a multicriteria approach seems to be convincing and capable of addressing all the dimensions involved in the problem. Among the positive aspects, the recent attention given to this decision problem deserves to be pointed out as well the awareness about potential impacts of choices on the environment, the whole society and also the general well-being, thus stressing that the importance for DMs for the implementation of tools or decision support systems able to address this kind of complex choices.

References

Abdullahi S, Mahmud AR, Pradhan B (2014) Spatial modelling of site suitability assessment for hospitals using geographical information system-based multicriteria approach at Qazvin city, Iran. Geocarto Int 29:164–184

Azzopardi Muscat N, Brambilla A, Caracci F, Capolongo S (2020) Synergies in design and health. The role of architects and urban health planners in tackling key contemporary public health challenges. Acta Biomed 91(3):9–20. https://doi.org/10.23750/abm.v91i3-S.9414

Bassi A, Ottone C, Dell'Ovo M (2019) I Criteri Ambientali Minimi nel progetto di architettura. Trade-off tra sostenibilità ambientale, economica e sociale. Valori e Valutazioni 22:35–45

Beheshtifar S, Alimoahmmadi A (2015) A multiobjective optimization approach for location-allocation of clinics. Int Trans Op Res 22:313–328

Brambilla A, Capolongo S (2019) Healthy and sustainable hospital evaluation—a review of POE tools for hospital assessment in an evidence-based design framework. Buildings 9(4):76

Brans JP, Gallo G (2007) Ethics in OR/MS: past, present and future. Ann Oper Res 153(1):165–178

BRE Global Ltd (2010) Breeam healthcare. BRE, Watford

Buffoli M, Capolongo S, di Noia M, Gherardi G, Gola M (2015) Healthcare sustainability evaluation systems. In: Capolongo S, Bottero MC, Buffoli M, Lettieri E (eds) Improving sustainability during hospital design and operation: a multidisciplinary evaluation tool. Springer briefs in applied sciences and technology. Cham, Switzerland, pp 23–30. https://doi.org/10.1007/978-3-319-140 36-0_3

Buffoli M, Nachiero D, Capolongo S (2012) Flexible healthcare structures: analysis and evaluation of possible strategies and technologies. Ann Ig 24(6):543–552

Burkey ML (2012) Decomposing geographic accessibility into component parts: methods and an application to hospitals. Ann Reg Sci 48:783–800

Burkey ML, Bhadury J, Eiselt HA (2012) A location-based comparison of health care services in four U.S. states with efficiency and equity. Socio-Econ Plann Sci 46:157–163

Capasso L, Faggioli A, Rebecchi A, Capolongo S, Gaeta M, Appolloni L, De AM, D'Alessandro D (2018). Hygienic and sanitary aspects in urban planning: Contradiction in national and local urban legislation regarding public health. Epidemiol e prevenzione 42(1):60–64

Capolongo S (2016) Social aspects and well-being for improving healing processes' effectiveness. Ann Ist Super Sanità 52(1):11–14

Capolongo S, Buffoli M, Brambilla A, Rebecchi, A (2020) Healthy urban planning and design strategies to improve urban quality and attractiveness of places. TECHNE 19:271–279. https://doi.org/10.13128/techne-7837

Cheng EWL, Li H (2004) Exploring quantitative methods for project location selection. Build Environ 39:1467–1476

Cheng EWL, Li H, Yu L (2007) A GIS approach to shopping mall location selection. Build Environ 42:884–892

Chiu JE, Tsai HH (2013) Applying analytic hierarchy process to select optimal expansion of hospital location: the case of a regional teaching hospital in Yunlin. In: 2013 10th international conference on service systems and service management. IEEE, pp 603–606

Daskin MS, Dean LK (2005) Location of health care facilities. In: Operations research and health care. Springer, Boston, pp 43–76

deAlmeida AT, Cavalcante CAV, Alencar MH, Ferreira RJP, deAlmeida-Filho AT, Garcez TV (2015) Multiobjective and multicriteria decision processes and methods. In: Multicriteria and multiobjective models for risk, reliability and maintenance decision analysis. Springer, Cham, pp 23–87

Dell'Ovo M, Capolongo S (2016) Architectures for health: between historical contexts and suburban areas. Tool to support location strategies. TECHNE-J Technol Architect Environ, 269–276

Dell'Ovo M, Capolongo S, Oppio A (2018a) Combining spatial analysis with MCDA for the siting of healthcare facilities. Land Use Policy 76:634–644

Dell'Ovo M, Frej EA, Oppio A, Capolongo S, Morais DC, de Almeida AT (2018b) FITradeoff method for the location of healthcare facilities based on multiple stakeholders' preferences. In: International conference on group decision and negotiation. Springer, Cham, pp 97–112

Dell'Ovo M, Frej EA, Oppio A, Capolongo S, Morais DC, de Almeida AT (2017) Multicriteria decision making for healthcare facilities location with visualization based on FITradeoff method. In: International conference on decision support system technology. Springer, Cham, pp 32–44. https://doi.org/10.1007/978-3-319-57487-5_3

Du G, Sun C (2015) Location planning problem of service centers for sustainable home healthcare: evidence from the empirical analysis of Shanghai. Sustainability 7:15812–15832

Faruque LI, Ayyalasomayajula B, Pelletier R, Klarenbach S, Hemmelgarn BR, Tonelli M (2012) Spatial analysis to locate new clinics for diabetic kidney patients in the underserved communities in Alberta. Nephrol Dial Transpl 27(11):4102–4109

Fredrickson JW (1986) The strategic decision process and organizational structure. Acad Manag Rev 11(2):280–297

Gimpel A, Stelzenmüller V, Grote B, Buck BH, Floeter J, Núñez-Riboni I, Pogoda B, Temming A (2015) A GIS modelling framework to evaluate marine spatial planning scenarios: co-location of offshore wind farms and aquaculture in the German EEZ. Mar Policy 55:102–115

Hammond JS, Keeney RL, Raiffa H (1999) Smart choices. Harvard Business School Press, Boston

ITACA (2011) Protocollo Itaca per la valutazione della qualità energetica ed ambientale di un edificio. ITACA, Rome

Japan Sustainable Building Consortium (2006) Comprehensive assessment system for built environment efficiency. Japan sustainable building consortium–[viewed on 20 Jan 2011]: Available on the Internet: http://www.ibec.or.jp/CASBEE/english/theassessment-method-employed-by-casbee

Kahraman C, Ruan D, Dogan I (2003a) Fuzzy group decision-making for facility location selection. Inf Sci 157:135–153

Kahraman S, Findikli N, Berkil H, Bakircioglu E, Donmez E, Sertyel S, Biricik A (2003b) Results of preimplantation genetic diagnosis in patients with Klinefelter's syndrome. Reprod BioMed Online 7(3):346–352

Keeney RL (1992) Value focused thinking. Harvard University Press, Cambridge

Keeney RL, Raiffa H (1993) Decisions with multiple objectives: preferences and value trade-offs. Cambridge University Press, Cambridge

Keeney RL (2004) Make better decision makers. Decis Anal 1(4):193–204

Kersten GE, Szpakowicz S (1994) Decision making and decision aiding: defining the process, it representations, and support. Group Decis Negot 3(2):237–261

Kim JI, Senaratna DM, Ruza J, Kam C, Ng S (2015) Feasibility study on an evidence-based decision-support system for hospital site selection for an aging population. Sustainability 7(7):2730–2744

Kumar S, Bansal VK (2016) A GIS-based methodology for safe site selection of a building in a hilly region. Front Architect Res 5(1):39–51. https://doi.org/10.1016/j.foar.2016.01.001

Lahtinen J, Salonen M, Toivonen T (2013) Facility allocation strategies and the sustainability of service delivery: modelling library patronage patterns and their related CO_2-emissions. Appl Geogr 44:43–52

Latinopoulos D, Kechagia K (2015) A GIS-based multi-criteria evaluation for wind farm site selection. A regional scale application in Greece. Renew Energy 78:550–560

Lee KS, Moon KJ (2014) Hospital distribution in a metropolitan city: assessment by a geographical information system grid modelling approach. Geospat Health 8:537–544

Lenzi A, Capolongo S, Ricciardi W, Signorelli C, Napier D, Rebecchi A, Spinato C (2020) New competences to manage urban health: health city manager core curriculum. Acta Biomed 91(3):21–28. https://doi.org/10.23750/abm.v91i3-S.9430

Li S, Cai H, Kamat VR (2015) Uncertainty-aware geospatial system for mapping and visualizing underground utilities. Autom Constr 53:105–119

Murad AA (2005) Using GIS for planning public general hospitals at Jeddah City. Env Des Sci 3:3–22

Murad AA (2007) Creating a GIS application for health services at Jeddah city. Comput Biol Med 37(6):879–889

Neutens T, Delafontaine M, Scott DM, De Maeyer P (2012) A GIS-based method to identify spatiotemporal gaps in public service delivery. Appl Geogr 32(2):253–264

Noon CE, Hankins CT (2001) Spatial data visualization in healthcare: supporting a facility location decision via GIS-based market analysis. In: Proceedings of the 34th Hawaii international conference on system sciences

Noorollahi Y, Yousefi H, Mohammadi M (2016) Multi-criteria decision support system for wind farm site selection using GIS. Sustain Energy Technol Assess 13:38–50

Oppio A, Buffoli M, Dell'Ovo M, Capolongo S (2016) Addressing decisions about new hospitals' siting: a multidimensional evaluation approach. Ann dell'Istituto superiore di sanità 52(1):78–87

Ormerod RJ, Ulrich W (2013) Operational research and ethics: a literature review

Owen SH, Daskin MS (1998) Strategic facility location: a review. Eur J Oper Res 111:423–447

Shariff SR, Moin NH, Omar M (2012) Location allocation modeling for healthcare facility planning in Malaysia. Comput Ind Eng 62(4):1000–1010

Simon HA (1979) Information processing models of cognition. Annu Rev Psychol 30(1):363–396

Soltani A, Marandi EZ (2011) Hospital site selection using two-stage fuzzy multi-criteria decision making process. J Urban Environ Eng 5(1):32–43

USGBC (2011) LEED for healthcare. USGBC, Washington

Vahidnia MH, Alesheikh AA, Alimohammadi A (2009) Hospital site selection using fuzzy AHP and its derivatives. J Environ Manage 90(10):3048–3056

Wang X, Hu P, Zhu Y (2016) Location choice of Chinese urban fringe residents on employment, housing, and urban services: a case study of Nanjing. Front Architect Res 5(1):27–38

Wu CR, Lin CT, Chen HC (2007) Optimal selection of location for Taiwanese hospitals to ensure a competitive advantage by using the analytic hierarchy process and sensitivity analysis. Build Environ 42(3):1431–1444

Wu WH, Lin CT, Peng KH, Huang CC (2012) Applying hierarchical grey relation clustering analysis to geographical information systems—a case study of the hospitals in Taipei City. Expert Syst Appl 39(8):7247–7254

Yang Y, Wong KK, Wang T (2012) How do hotels choose their location? Evidence from hotels in Beijing. Int J Hospital Manage 31(3):675–685

Zhang P, Ren X, Zhang Q, He J, Chen Y (2015) Spatial analysis of rural medical facilities using Huff Model: a case study of Lankao County, Henan Province. Int J Smart Home 9:161–168

Chapter 2
Structuring the Decision Problem. A Spatial Multi-methodological Approach

Abstract Multicriteria Decision Analysis (MCDA) allows to establish preferences about multiple options by considering both qualitative and quantitative data. MCDA is structured in main stages aimed at supporting Decision-Makers (DMs) to establish the decision context, elicit main objectives, explore potential decisions by assessing their performances and to take a final decision. Within the decision context concerning the location of healthcare facilities, four main steps of the stages characterizing the MCDA have been investigated given their relevance in affecting the decision problem, namely the presence of a variety of stakeholders, the definition of a consistent set of criteria, the choice of the criteria weight elicitation procedure and the selection of an aggregation procedure. The purpose of the contribution is to present some prominent existing methodologies aimed at developing the four steps identified in order to detect strengths and weaknesses. Moreover, given the nature of the decision problem, main features of the spatial analysis will be discussed.

2.1 Introduction

Considering the definition of Decision Aiding given by Roy (1985, 2005), in the context of complex decision processes, it is rare to solve a problem by a single criterion. By taking into account only one point of view and its underlying single criterion (mono-criterion approach) will lead to neglect important aspects. When the problem to be solved concerns the location of healthcare facilities, it is more effective the use of a multicriteria approach rather than a monodimensional one. Solving a problem by including different points of view and the several aspects involved, will facilitate the debate and will bring to light unexpected considerations for the moment when a decision has to be taken (Roy 1985, 2005). The Multicriteria Decision Analysis (MCDA) is defined both as an approach and a set of techniques able to provide an overall rank of alternatives (DCLG 2009) able to establish preferences about multiple options. According to the data available, the type of decisions, different procedures can be chosen and applied.

It considers at the same time qualitative and quantitative data and has the important task of providing Decision-Makers (DMs) with specific models able to solve real situations, by identifying a rational path for a desirable outcome. Every analysis is unique, but they generally have some common features (Hwang and Yoon 1981):

- a variety of attributes;
- conflicting criteria;
- disparate measurement units;
- identification of the best alternative to solve the initial issue.

The main achievement of this approach is to define criteria upon which the final decisions are taken. MCDA, as an example, can determine the potential of an intervention in the territory and unlike evaluation techniques that are purely based on monetary assessments for suitability analysis, MCDA responds with more adequate tools, by combining criteria weighted on stakeholders' priorities (Faroldi et al. 2019; Capolongo et al. 2015a; Malczewski and Rinner 2015; Ferretti 2012). The final purpose of MCDA is not to suggest the perfect decision, but to explore multiple potential decisions by breaking the problem into more comprehensive levels (decision tree, value tree). Breaking down the problem helps to understand the value of each single level and to not disregard its important aspects. (DCLG 2009).

In agreement with Roy (1985), MCDA:

- "It facilitates taking account of a broad spectrum of points of view liable to structure a decision making process for all the relevant actors;
- By making a family of criteria explicit, it preserves the original concrete meaning of the corresponding evaluations for each actor, without resorting to any fabricated conversion;
- It clears the way for a discussion on the respective roles that each criterion may play during the decision aiding process, e.g., weight, veto, aspiration level, rejection level".

According to Keeney and Raiffa (1976), MCDA techniques provide a support for DMs to develop coherent preferences since they are framed step by step with the investigation of the problem and not a priori. Once the problem is well known and the DM feels confident about it, the decision can be taken. Main stages of the MCDA consider (DCLG 2009):

- *Establishing the decision context*: implies a deep knowledge about the stakeholders affected by the problem and able to influence it. Moreover, it is important to know the state of the art of the problem to solve in order to have clear in mind strengths, weaknesses, opportunities and threats;
- *Identifying objectives and criteria*: allows to understand which are the main objectives to achieve and to hierarchically structure the problem into levels, namely decision tree or value tree, composed by criteria and sub-criteria aimed at supporting the problem resolution;
- *Identifying alternatives*: considers the identification (when a set of alternatives is already given) or the definition (result of the combination of different strategies able to solve the problem) of a set of possible options;

- *Assessing the performance of alternatives*: implies the assessment of alternatives' performance in relation to the set of criteria and sub-criteria framed and the specific indicators identified to measure them (qualitative or quantitative) for the comparison and the final evaluation. Commonly a performance matrix is used to explore alternatives;
- *Criteria weight elicitation*: allows the assignment of a different importance to the set criteria and sub-criteria defined since they can have a difference influence in achieving the final aim;
- *Aggregation of weights and scores*: underlies the standardization, where all the measurement units previously selected become uniform and the scores lose their dimension. Once all the performances have been standardized, they can be aggregated with the weights assigned;
- *Result*: allows to visualize the overall rank and also the partial results obtained, in order to evaluate the performance of the alternatives under each aspect and level of the hierarchy in order to take a more conscious decision;
- *Sensitivity analysis*: aims at validating the result obtained and at understanding its internal robustness.

Within this context, according to the literature analysed on this topic (see Chap. 1), it is possible to recognize four main problems capable of affecting the decision problem concerning the location of healthcare facilities and present in the main stages previously described. The first involves the variety of stakeholders with their own interests to fulfil and power to prioritize (i); the second one concerns the set of criteria considered able to describe the complexity inherent in the problem (ii), the third one involves the choice of the criteria weight elicitation procedure (iii) and the fourth one concerns the selection of a methodology for the aggregation of data obtained by the analysis (iv). These research questions have been widely analysed by scholars but deserve to be better investigated within the location of healthcare facilities (Dell'Ovo et al. 2016, Dell'Ovo et al. 2018a, b).

The contribution will propose an overview of some prominent existing methodologies aimed at developing the four stages identified as the more critical in this decision context (Sect. 2.2). Given the nature of the problem it will be explored the role played by Geographic Information Systems (GIS) (Sect. 2.3) and the main achievements will be synthetized in the conclusions (Sect. 2.4).

2.2 How to Structure the Decision Problem

Within this section, the main important methodologies developed in order to identify the stakeholders, a consistent set of criteria, suitable weights assignment methods and aggregation procedures will be described. This phase becomes fundamental in order to show how to combine different techniques for facing a complex problem such as the location of healthcare facilities.

2.2.1 Stakeholder Analysis

Facing the location of healthcare facilities requires the participation of different categories of actors, being the final decision responsive of several impacts on the whole society, the natural environment as well as the built environment. In Italy, Regions have the responsibility of promoting health policies, managing resources and satisfying the needs of a wide range of actors. There are political actors that own the power to stop proposals or to support them, then there are healthcare companies, representing the private sector and their own interests but promoting at the same time public health (Capolongo et al. 2015b). Other actors involved in the process could be: users, health workers, regular patients, helpers etc.

According to Dente (2014) there are mainly three typology of actors involved in Decision-Making processes:

- Stakeholders: actors affected by the decision or who have the possibility to influence it;
- Decision-Makers (DM): are responsible of taking the decision;
- Analyst: supports the DM in taking decisions, allow to understand which knowledges have to be used and to elicit DM objectives;

Actors are defined as those who perform the relevant actions and takes important decision. Yang (2014) provides an overview of the main definition of stakeholder analysis by investigating the existing literature. According to their description given by Dente (2014) "the outcome of a decisional process depends on the actors" and it is possible to identify people involved into the process once the problem has been structured. This mechanism allows to specify their role and in particular whose interests are involved. In "Understanding policy decisions" Dente (2014), besides the analysis of decisional situations, suggests operational recommendation in order to perform a complete stakeholder analysis. For example, we could run into 'methodological constitutionalism' in case only actors that are legally responsible for implementing a project are considered, in fact he underlines the importance of the 'empirical constitutionalism' where real world situations are completely understood. Another important consideration comes to light in case of group of actors. Dente clarifies about the presence of two different typologies of groups of actors, in the first case they act for a common good, the interaction among them is stable, no pressure, everyone is free to express his/her own opinion. On the other hand, there could be groups of actors with different expectations and interests, in this case it is not possible to consider them collectively and aggregate their opinions but they should be analysed individually, trying to understand the actors' behaviour and potential conflicts.

Understanding actors involved into decision processes plays a crucial role and according to their power to influence the final choice and their needs, the criteria and the weights can radically change.

Moreover, if the problem under investigation involves the interaction with multiple actors with specific influences on the decision process, it may become more and more complex. In fact, how it has been well described by de Almeida and Wachowicz

(2017), decisions with multiple DMs are more challenging compared to individual decisions since there is a co-existence both of conflicting objectives and different viewpoints, preferences and aspirations.

The aim of the Stakeholder Analysis is to understand how to classify actors involved and possible ways to interact with them; in which phases of the decision process they can intervene and in case of multi-stakeholders how to define their influences for achieving the final aim.

There are different methodologies able to understand which are the main categories of stakeholders involved and their relation.

The literature mainly divides actors into categories in accordance with the resources they control (Dente 2014). Political resources are related to the amount of consensus an actor is able to obtain. Economic and Financial resources consist in the ability to modify other actors' behaviour through the use of money. Legal resources consider the behaviour of actors that are legitimate by their administrative or legislative authority. Cognitive resources mean to own specific knowledges and experiences about the decision problem. Each actor behaves and acts in accordance with the type of resources previously defined (Dell'Ovo et al. 2017, 2018b). In detail, five different categories of stakeholder can be recognised:

- Political actors, who represent citizens, enjoys significant consensus;
- Bureaucratic actors, who has formal competence to intervene;
- Actors with special interests, who sustains costs or benefits from decision-making process. The choice between the alternatives directly affects their interests;
- Actors with general interests, who represents people and/or interests not able to defend themselves (future generation, fauna, consumers, etc.);
- Experts, who has specific knowledge in order to define the decision-making process and to identify suitable alternatives to solve it.

Political actors move political resources, bureaucratic actors move legal resources; actors with special interests move economic resources; actors with a general interest move both legal resources and cognitive resources; and experts move cognitive resources. Anyway one actor can be classified across different categories according to his role and the kind of resources he moves.

A further analysis concerns the investigation of the level of interest and power of the stakeholder previously defined (Mendelow 1981).

The level of influence represents how much an actor is able to influence the achievement of an objective in the decisional process while the level of interest represents how much the decisional process can affect the objectives- activities of the stakeholder. Based on the relevance of their interest and influence, the actors can be classfied by a matrix into the following categories (see Fig. 2.1):

- Key players: strong capacity of intervention on decisions;
- Keep satisfied: element of pressure, able to influence public opinion;
- Keep informed: often coincides with the beneficiaries of the program/plan/project;
- Minimal effort: affected by decisions.

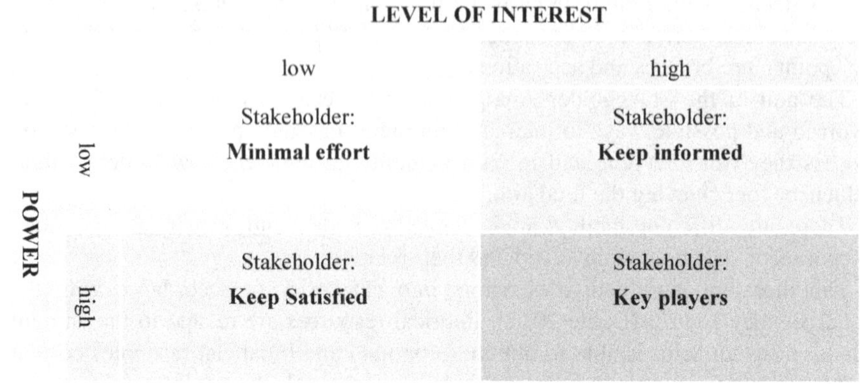

LEVEL OF INTEREST

	low	high
low	Stakeholder: **Minimal effort**	Stakeholder: **Keep informed**
high	Stakeholder: **Keep Satisfied**	Stakeholder: **Key players**

(POWER — vertical axis label on the left)

Fig. 2.1 Power/Interest matrix. Adapted from Mendelow (1981)

Another methodology for the stakeholder analysis is the Stakeholder Circle methodology developed by Bourne (2005). It provides a tool able to identify and prioritize key stakeholders to define a strategy of engagement and communication and to ensure that the needs and expectations of these stakeholders are understood and managed according to a process divided into five key steps (identification, identification of priorities, visualization, involvement, monitoring). Three attributes are used to evaluate the relative importance of the stakeholders:

- power: do their power significantly influence the work or the outcomes of the project?
- proximity: are they closely associated or relatively remote from the work of the project?
- urgency: are they prepared to go to any length to achieve their outcomes?

The Stakeholder Circle includes two key elements:

- the concentric circles that indicate the distance of the stakeholders from the program/plan/project under investigation;
- the patterns that indicate the homogeneity of each category of stakeholder.

The size and the area of each segment indicate the scale and the influence, while the radial depth the ability of the stakeholders to condition a project (see Fig. 2.2).

Moreover, the Social Network Analysis (Dente 2014) considers the size and ownership of decision-making networks. In fact, to understand who the actors are, what kinds of objectives they pursue, what resources they exchange, what action logics follow and what role they play in the decision-making process, it is important to ask whether and how the structure of interaction is a further and distinct causal factor. To measure the density of the relation among stakeholders, different indices have been developed. The Density index deals with the property of the network of considering the amount of relationships established between the actors involved in

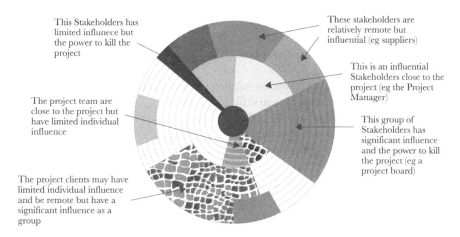

This Stakeholders has limited influnece but the power to kill the project

These stakeholders are relatively remote but influential (eg suppliers)

This is an influential Stakeholders close to the project (eg the Project Manager)

The project team are close to the project but have limited individual influence

This group of Stakeholders has significant influence and the power to kill the project (eg a project board)

The project clients may have limited individual influence and be remote but have a significant influence as a group

Fig. 2.2 Stakeholder circle. Adapted from Bourne (2005)

the process. The Centrality index takes into consideration the fact that one or a few actors "monopolize" relations among other participants.

Dealing with complex systems, the adoption of a participatory paradigm is suggested, since different competences are required to solve the problem and the final strategy adopted should reach a general consensus and stimulate an active debate among "players".

2.2.2 Criteria Definition

Criteria are those "element" able to describe the problem in all its part. They can be both qualitative and quantitative according to the specific characteristic they are describing. There is not a common set of criteria universally shared by the community able to solve any kind of problem, each situation could be described by a specific set of criteria and then measured in a different way. In fact, once elicited suitable criteria, it is also important to define in which way they should be assessed.

The definition of an appropriate set of criteria should be consecutive to the definition of the decision problem, it cannot be identified a priori since it is a strategic framework able to describe the issues considered and to find possible solutions. Criteria can help to compare alternatives or to evaluate one single alternative, and according to their use and the unit of measurement selected for each criterion, the result could change. According to the indicator chosen the performance of one criterion could be maximized (the higher performance is the best) or minimized (the lower performance is the best).

Criteria should have the following characteristics (Roy 1985, 2005):

- Systemic: able to take into consideration all the aspect of the decision problem reflecting its essential characteristics;
- Consistency: avoiding redundancy and to be consistent with the objectives elicited by the DM;
- Independency: there should not be relation of dependency among criteria, each one should describe a specific and different aspect;
- Comparability: it is fundamental to define criteria that can be compared otherwise it is impossible to go on with the evaluation;
- Measurability: each criterion has to be described by an indicator able to measure it.

Criteria represent our objectives, what we want to achieve for improving the business as usual scenario (Keeney 1992). Understanding what the DM and other stakeholders consider important, why and in which measure will support in taking a decision and in framing a consistent evaluation framework. It becomes a priority to identify the issues of interest and the boundaries of the decision, after that, objectives and criteria are typically laid out in a hierarchy (decision tree, value tree, etc.) (Bottero et al. 2019).

In addition to the characteristics previously defined:

- Criteria must relate to specific objectives;
- Criteria can be dictated or case specific;
- Criteria typically evolve during the analysis.

The purpose is to define a set of criteria useful to evaluate the performance of a set of alternatives. As already explained, usually criteria are organised in hierarchy levels (Saaty 1980) and aggregated in macro-areas to facilitate the comparison. Once defined the final objective of the evaluation/what we are looking for/what we want to achieve, it is possible to create our value tree composed by levels. The first level is devoted to the final aim, then the set of criteria (macro-area) is framed further divided in sub-criteria specified by indicators (see Fig. 2.3).

Different approaches have been developed in previous researches in order to identify a suitable set of criteria, within this contribution some of them will be presented.

Keeney has developed the Value-Focused Thinking (VFT) approach in 1992. The VFT is an approach able to guide DMs across all the phases of the process, from the elicitation of the objectives to the identification of the most suitable alternatives. It is composed by four main steps (Keeney 1992):

- identifying objectives;
- structuring objectives;
- creating alternatives;
- identifying decision opportunities.

These four steps could be processed one after the other or individually, but anyway all of them brings to better decisions. By focusing the attention on the first two steps, the approach is able to elicit stakeholders' values, that are transformed in objectives,

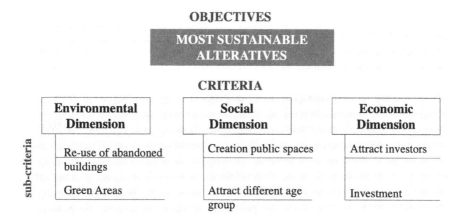

Fig. 2.3 Example of a value tree

then structured in a value-tree. In particular "values are the reason to spend any time thinking about the decision" (Keeney 1992). Once values have been recognised, they can be easily stated as objectives. A value is anything a DM cares about in a decision, while an objective is a value stated in a verb-object format. The last phase to process in order to obtain a value-tree is the comprehension of the relationships and the role of objectives. It is necessary to recognize Fundamental Objectives from Means Objectives:

- Fundamental Objectives: an objective that defines a basic reason for caring about a decision;
- Means Objectives: an objective whose importance stems from its influence on achieving other objectives.

While most of the decision-making processes are based on alternatives already given—Alternative Focused Thinking (AFT)—Keeney overturns this concept being the alternatives the results of a process and not anymore their starting point. Values come first.

This kind of approach can help to understand why we care about a decision to make, to prioritize our objectives and finally to state them as criteria (Dell'Ovo and Oppio 2019).

Another possible way to explicit criteria is well explained by Gamboa and Munda (2007), in a real case study related to the location of a windfarm. First of all, they identified stakeholders involved in the process, their scale of action and their position regarding the problem. After that, their needs and expectations have been elicited and transformed into criteria, in order to understand how to satisfy each stakeholder. More in detail, they have applied the Social Multi-Criteria Evaluation (SMCE) (Munda 2004) and modified the traditional relationship between DM and analyst by focusing on a social framework which becomes fundamental in the case of public choices. As already explained, the set of criteria is the translation of a stakeholder analysis where preferences of each social actor have been recognised.

This kind of approach can help in the case of public choices and participatory processes with multiple and conflictual stakeholders.

When a direct investigation of the stakeholders is not possible, the analysis can be based on the literature review. Through this process, it is possible to find out similar researches or case studies, whose information can be assumed as references. In this case it is important to perform the analysis through databases (e.g. Scopus, Web of Science, etc.) that allow the use of keywords to identify scientific papers according to the research framed. Papers can be then further classified according to the title and the abstract in order to understand their relevance. Once selected the most pertinent ones, the investigation is carried on with a deep analysis that consider for each article the methodology applied to solve the problem, the decision context, the scale, the criteria involved, as well as the tools and software, etc. Deeper is the analysis and deeper is the knowledge about the state of the art. After this classification it is possible to proceed by identifying the most frequently used and the more meaningful criteria for the case under analysis.

This kind of approach can address the definition of a comprehensive framework of criteria.

2.2.3 Criteria Weight Elicitation

Another key concept to consider is the criteria weight elicitation. Many scholars are investigating this issue and according to Riabacke et al. (2012) is still defined challenging, not solved and "there are still no generally accepted method available", even if it is a problem faced by many research areas. Someone argues because of the difficulty of giving precisely all parameters value, for the cognitive limitation of the DM (Yang et al. 2014) or maybe the not awareness about the problem and then the incoherence (Figueira and Roy 2002).

This phase is particularly influential, because according to the procedure used to elicit criteria the result could change and the inconsistence could grow or be reduced. In fact, one of the main issue related to the weight assignment regards the methods applied and also the stakeholder(s) interviewed.

The location of healthcare facilities could be a good example for analysing this step, as it is a process that should consider many different stakeholders (Burkey et al. 2012) with different expectations and preferences. The methodology selected should be suitable according to the number and the type of actors. In fact, since in public decisions it is also important to consider the opinion of non-expert stakeholders (as the citizens), it becomes fundamental the use of approaches easy to understand and able to catch their impression without conditioning them.

The aim of this step is to assign different influence to the criteria involved in the problem since, according also to the context (social, economic, etc.), they can have a different importance in achieving the final purpose (Assumma et al. 2019). Assigning a weight is already a first moment where a judgment become explicit. In fact, here stakeholders have to show their preferences. In this phase the role of the analyst is fundamental since he should not influence the actors involved and should be able to explain how their opinion could drastically change the final result. Everyone should tend to strength his own interests and to carry on his needs, this is the reason why in public decision, in order to make a decision able to satisfy the demands of the whole society, it is strategic to apply a methodology able to consider different points of view (Capolongo et al. 2019).

Since there is not a general accepted method (Riabacke et al. 2012), within this context, it is possible to appreciate an overview of the most prominent weight elicitation methods framed by Riabacke et al. (2012), in "State of the Art Prescriptive Criteria Weight Elicitation". With respect to the original one, it has been added a new method recently developed—the FITradeoff (de Almeida et al. 2016)—for the elicitation procedure (see Table 2.1). Moreover, in the same paper, there is a summary of the assessment procedures during the extraction step of the weight elicitation methods (see Table 2.2).

2.2.4 Aggregation Procedures

A DM may have one of the following questions with respect to the set of available alternatives under evaluation:

- How to choose the most suitable action?
- How to classify actions into predefined decision classes?
- How to order actions from the best to the worst?

According to the aggregation procedure it is possible to answer to these questions.

Given the phases previously explained, the aggregation can be seen as the conclusive one, since the final evaluation is able to take into consideration both the alternatives' performance and the weights assigned. Moreover, according to the methodology selected, different results could be obtained (Greco et al. 1999, 2001, 2004, 2009; Ferretti and Montibeller 2016):

- Choice problem (optimization): allows to select a restricted subset of suitable alternatives able to satisfy the objective of the decision problem;
- Classification problem (sorting): allows to assign each alternative into categories that correspond to a set of classes. In the classification problem classes are nominal, in sorting problem classes are ordered;
- Ordering problem (ranking): allows to rank alternative by order of preference. The regulation can be given in partial or complete ranking of actions.

According to this classification, the methodology to apply has to consider the aim of the research, the set of criteria and the expected result. There are mainly three aggregation procedures:

Table 2.1 Weight elicitation methods. Adapted from Riabacke et al. (2012)

Weight elicitation method	Source	Assessment	Extraction input	Min. no. of judgments	Representation	Interpretation
Direct rating	Comparison with PA (Bottomley et al. 2000)	Cardinal joint procedure	Precise	N	Point estimates	Normalized criteria weights
Point allocation	Comparison with DR (Bottomley et al. 2000)	Cardinal joint procedure	Precise	N	Point estimates	Normalized criteria weights
SMART	(Edwards 1977)	Cardinal joint procedure	Precise	N	Point estimates	Normalized criteria weights
SWING	(von Winterfeldt and Edwards 1986)	Cardinal Joint procedure	Precise	N	Point estimates	Normalized criteria weights
Trade-off	(Keeney 1992; Keeney and Raiffa 1976)	Cardinal Pairwise procedure	Precise	$N * (N - 1)$ with consistency check	Point estimates Relative between pairs criteria	(Combined) normalized criteria weights
Rank-order methods	Rank Reciprocal (RR) weights (Stillwell et al.,1981) Centrois (ROC) weights (Barron 1992) SMARTER (SMART Exploiting Ranks) (Edwards and Barron 1994)	Ordinal joint procedure	Rank-order	N	Comparative statement	Surrogate criteria weights (translated using a conversion method)
AHP	(Saaty 1980)	Cardinal Pairwise procedure	Semantic	$N * (N - 1)$ with consistency check	Semantic estimates Relative between pairs criteria	(Combined) surrogate criteria weights (translated from semantic to exact numerical)

(continued)

Table 2.1 (continued)

Weight elicitation method	Source	Assessment	Extraction input	Min. no. of judgments	Representation	Interpretation
CROC	(Riabacke et al. 2009)	Ordinal and Cardinal Joint procedure	Rank-order and imprecise cardinal relation information	N (>N with cardinal input)	Comparative statement + Imprecise cardinal relation information	Surrogate (centroid) criteria weights
Interval methods	(Walley 1991; Danielson and Ekenberg, 2007)	Normally, a generalized ratio-weight procedure	Interval endpoints (precise)	2 * (min. no. of judgments in employed ratio-weight procedure)	Intervals	Interval criteria weight
FITrade-off	(de Almeida et al. 2016)	Cardinal Pairwise procedure	Strict preference procedure	Rank order of criteria weights (N)	Bounds for criteria weights	Weight space (set of feasible weight vectors)

Table 2.2 Extraction step of the weight elicitation methods. Adapted from Riabacke et al. (2012)

Weight elicitation method	Extraction assessment procedure
Direct rating	Rate each criterion on a 0–100 scale
Point allocation	Distribute 100 points among the criteria
SMART	(1) Identify the least important criterion, assign 10 points to it (2) Rate the remaining criteria relative to the least important one
SWING	(1) Consider all criteria at their worst consequence level (2) Identify the criterion most important to change from worst to best level, assign 100 points to it (3) Continue with steps 1 and 2 with the remaining criteria, rate relative to the most important
Trade-off	Judge criteria in pairs (1) Make a choice between two alternatives alt.1: the best consequence level of the first criterion and the worst of the second alt.2: the worst consequence level of the first criterion and the best of the second (2) State how much the decision-maker is willing to give up on the most important criterion in order to change the other one to its best level (3) Continue with steps 1 and 2 with the remaining criteria
Rank-order methods	Ordinal statements of criteria importance, that is rank all criteria from the most important to the least important
AHP	Use a systematic pairwise comparison approach in determining preferences (1) Make a choice between two criteria to determine which is the most important (2) State how much more important the criterion identified in step 1 is in comparison to the second criterion using a semantic scale to express strength of preference (3) Continue with steps 1 and 2 with the remaining criteria
CROC	(1) Rank all criteria from the most to the least important (2) The most important criterion is given 100 points. The decision-maker is asked to express the importance of the least important criterion in relation to the most important (3) Adjust the distances between the criteria on an analogue visual scale to express the cardinal importance information between the criteria
Interval methods	Generalized ratio weight procedures which employ interval judgments to represent imprecision during extraction instead of point estimates, as in for example, interval SMART/SWING
FITrade-off	Similar to the trade-off procedure, but the DM answers only strict preference questions

- Compensatory;
- Partially-compensatory;
- Non-compensatory.

Compensatory method
Inside this cluster different methods have been developed (e.g. Multiple Attributes methods (MAUT); Weighted average; AHP/ANP). They are described by value functions able to approach deterministic consequences and aggregated through analytical models by allowing trade-off (compensation) between criteria. A weak performance obtained against one criterion can be compensated by a good one obtained against another criterion (DCLG 2009). Here criteria have to be independent in terms of preferences. The final aim is to provide an efficient solution, meant as a balance among the set of criteria as well as a good compromise in relation to the other alternatives.

Partially-compensatory methods
Inside this classification different methods have been developed (e.g. PROMETHEE; NAIADE; REGIME; ELECTRE). Here the DM is further interviewed to obtain complement information and supplementary judgement. Outranking (Roy 1974) relations are built according to a series of pairwise comparison between every pair of options being considered. These methods involve usually two phases, during the first one outranking relations are defined and in the second one an overall preference ranking is suggested (Bouyssou 2008).

Non-compensatory method
In this context it is possible to mention the Dominance method where decision rules are stated according to the format "if ... then ...". By applying this methodology, it is possible to take into consideration the most complex interaction among criteria. The decision rules are the result of the alternatives' performance and the general overall evaluation obtained by them, that means that a causal relationship is present. The Dominance method allows through the induced rules to convey a useful knowledge as they depict some regularities observed in the analysed data set, that can be also used to classify new alternatives for which the decision is still to be made (Pawlak 1982, 1991; Greco et al. 1999, 2001, 2009; Oppio et al. 2020).

More in general, the MCDA could be considered based on subjective judgments since (Greco et al. 1999, 2001, 2004, 2009):

- in real decision-making processes the borderline between what is feasible and what is not is not clear;
- the weights assignment is based on DM preferences;
- the interaction of the analyst can influence the DM;
- data collected could be imprecise and uncertain;
- mathematical models are not enough to define whether a decision is good or bad;
- subjectivity is a component of the decision aiding.

To overcome the subjectivity after a final rank has been obtained, it can be performed a sensitivity analysis able to validate and check the robustness of the result and to understand the stability of the result to changes of inputs.

2.3 Structuring the Spatial Problem

According also to the reviewed literature (see Chap. 1), locating healthcare facilities is a decision problem mostly characterized by spatial variables. Therefore, it is suggested the integration of spatial functions typical of GIS with the MCDA in order to get a wide and integrated knowledge about territory typical of Multicriteria-Spatial Decision Support Systems (MC- SDSS) by considering the spatial dimension of decision problems (Dell'Ovo et al. 2018a; Oppio et al. 2016).

Combining the two distinct domains of MCDA and GIS can provide new ways to face the decision problem. MCDA is considered by Roy (1985) a revolution in the field of the Operational Research whereas GIS comprehend a collection of tools able to collect, transform and display geographic data and preferences according to specific purposes (Burrough and McDonnell 1998). MCDA and GIS can both benefit from each other, improve their own strengths and expand the decision support domain (Densham and Goodchild 1989; Li et al. 2004; Ferretti 2012; Malczewski and Rinner 2015). By combining these two domains, the DM is able to analyse a problem structured according to a spatial multicriteria framework.

Since basic notion concerning the MCDA have been described in the introduction, this section is devoted to explain the potentialities of working with the GIS.

2.3.1 Geographic Information System (GIS)

Considering the definition given by the Encyclopaedia of Database Systems (Goodchild 2009), "Geographic Information Systems (GIS) is a computer application designed to perform a wide range of operations on geographic information." In fact, it is a mixture of tools and procedures that permit to store and process a large number of data, showing geographical information in a quite agile way. Nowadays with the evolution of technology and the increased awareness about the importance of open data, the amount of information available is massive. Moreover, the application of these "spatial tools" are found in different disciplines which issues analysed deal with "spatial phenomena" (Goodchild 2009). The benefits obtained by this method is the possibility to collect and analyse a huge quantity of geographical data and to conduct operations useful for the decision process, under a georeferenced environment.

Anyway, there are some concerns regarding this massive amount of data available:

- to select the correct information, with a high reliability;
- to avoid to be overwhelmed by an overload of data, by using those ones meaningful to criteria definition.

Typically, GIS is usually conceptualized as a collection of georeferenced layer, based on two different kind of data—raster and vector. The main difference among the two sets of data concern the number of properties a layer is able to collect. While raster-based layers are able to describe only a single property, vector-based layers could collect info regarding discrete objects with associated any number of attributes (Goodchild 2009).

Given this first overview, it is possible to conclude mentioning the fact that GIS software is constantly evolving towards the direction of "greater support for dynamics" (Goodchild 2009).

2.3.2 Multicriteria-Spatial Decision Support Systems (MC-SDSS)

Malczewski and Rinner (2015) in the book "Multicriteria Decision Analysis in Geographic Information Science" have deeply explained the origin of MC-SDSS and the link between GIS, spatial analysis and decision support systems:

- Spatial analysis concerns a set of techniques and models that operate with spatial processes and are directly dependent on the location features;
- Spatial decision supports born from the necessity to face complex spatial decision problems with the support of GIS.
- Spatial Decision Support System (SDSS) is a computer-based system developed to support DM in decision-making processes that consist in semi-structured spatial decision problems (Malczewski 1999), such as the location-allocation problems. In this system is fundamental the interaction with DM in order to improve the efficiency of operations and the coherence of the result. So far the GIS are not able to manage DM's preferences and judgments, for this purpose MCDA techniques are combined into GIS.
- MC-SDSS is defined as a class of SDSS based on the combination of GIS and MCDA, aimed to strength the potentials of GIS in the context of decision-making. MC-SDSS considers both geographic data and DM preferences as input that will result in maps as output. The maps resulted from the integration is not a decision in itself, but they can support the DM in taking a decision, by investigating pros and cons of each potential alternative, thus becoming a way for facilitating the discussion and the dialogue. One of the main achievement is the possibility to integrate data belonging to different fields.

Typically, MC-SDSS involves mainly three stages:

- standardization: first of all, it is important to define a set of criteria able to consider the spatial dimension of the problem and then it is necessary to transform them to comparable units through the standardization that is a mathematical representation of the human preferences (Keeney 1992). According to the software used and to the geo-operation applied to the original data (distance; density; etc.) it is possible to assign a dimensionless score (0–1; 0–10) to different performances;
- criteria weight elicitation: it assesses the relative importance of one criterion compared to the other criteria;
- aggregation rule: a dichotomic classifications can be defined as "compensatory versus non-compensatory, multiattribute versus multiobjective, discrete versus continues methods, and spatially implicit versus spatially explicit MCDA".[1] The weighted linear combination is an example of compensatory model since it allows the trade-off between criteria, while the Boolean overlay operations (conjunctive or disjunctive) is an example of non-compensatory model.

After a general overview of the main features of MC-SDSS, a literature review has been performed in order to outline the decision context of researches that explores the combined use of MCDA and GIS and to focus the attention on location problems.

According to Prasara-A and Gheewala (2017) and Moghadam et al. (2017) the analysis can be structured into four-stages:

1. "Literature search": databases can support this phase in order to provide documents and researches about the topic of interest;
2. "Screening process": keywords can better detail and narrow the analysis;
3. "Selection of literature": allow to identify papers more appropriate to the aim of the research according to the title; abstract; field; etc.
4. "Including literature": through the reading of the selected papers it is possible to collect information about how scholars have faced the decision problem under investigation.

By following the framework just described, in the first phase the Scopus database has been consulted by using a set of keyword as suggested by the second stage: "MCDA" or "Multi-Criteria Decision Analysis" and "GIS" or "Geographic Information System" and "Location". From the documents resulted (100), it is clear how the interest in combining MCDA and GIS is quite recent with a peak recorded in 2016. By the other two stages it has been possible to comprehend how many fields of research have covered this topic, mainly location problems related to the Energy demand, Risk assessment, Waste management, Geology and Urban planning (Gigović et al. 2017; Mat et al. 2016; Abdul-Mawjoud and Jamel 2016; Capolongo et al. 2016).

From these researches it has been moreover highlighted that:

---

[1]For the purpose of the research only the first classification will be explained, for further information please see "Malczewski and Rinner (2015). Multicriteria decision analysis in geographic information science. New York: Springer."

- only one document analysed is an Italian case study (Borgogno-Mondino et al. 2015);
- two paper are focused on investigating health issues (Varatharajan et al. 2017; Ho et al. 2015);
- only one paper is aimed to select a site for a new hospital (Vahidnia et al. 2009).

Given these premises a further investigation has been considered necessary to better understand the State of the Art, and to provide an overview of existing multi-methodological approaches supported by the GIS to locate healthcare facilities.

Considering again the four-stages approach previously explained:

- the literature review has been carried out using the Scopus database;
- the Screening phase has been performed by the use of the following keywords "hospital" or "healthcare facility"; "GIS" or "Geographic Information System" and "Location"; and removing "MCDA" or "Multi-Criteria Decision Analysis" since they have been considered too narrow.
- from the 425 documents found out, two sorts have been performed, the first by the title and the second by the abstract;
- 14 documents have been selected and then analysed.

The result of the literature review can be appreciated in Dell'Ovo et al. (2018a), to synthetize main important findings of the analysis developed, it is possible to conclude that GIS gave a huge support by the possibility to visualize immediately the outputs and to effectively communicate them. Moreover, when the methodology applied concerned the use of MCDA techniques, improvements have been detected (Vahidnia et al. 2009), since the combination helps to solve ill-structured problems where DMs does not have all the information and specification about the alternatives.

From the investigation have emerged:

- differences in the type of building/facility to locate;
- case by case approach and lack of a common methodology;
- combination of spatial visual system and multicriteria analysis;
- preferences for using Analytic Hierarchy Method (AHP) or the Analytic Network Process (ANP) to decompose the decision problem;
- lack of Italian case studies.

Finally, Vahidnia et al. (2009) clearly declared the necessity to develop an evaluation framework aimed to locate healthcare facilities based on spatial decision-making process.

2.4 Conclusions

This chapter has outlined some potential answers to the four more critical stages of the MCDA with respect to the location of healthcare facilities by showing different methodologies and how scholars have faced these issues (Bottero et al. 2015). First of all, a definition of MCDA has been given in order to better understand the general methodological approach and then the role played by the stakeholder analysis for an in-depth comprehension of the decision context has been investigated. Procedures able to identify and prioritize stakeholders' preferences have been explained. After that, the attention has been moved to the selection of a set of meaningful criteria, being criteria the expression of DM or stakeholders' values and objectives. Different criteria can have a different influence in achieving the overall goal, especially when weights are introduced. With the purpose of explaining the effects of criteria and weights, different methodologies have been presented. To finally rank the alternatives, an aggregation procedure has to be chosen according to the evaluation perspective and the available data.

Since the decision problem under investigation deals with spatial data, MC-SDSS, that combines MCDA and GIS, has been selected has the most adequate approach and then described. The abovementioned methodological steps have been explained both from a theoretical and an operational point of view and a literature review has been performed with the aim of (i) focusing the attention on the methodologies applied for the location of healthcare facilities; (ii) highlighting location problems already supported by the use of MC-SDSS; (iii) underlying how many scholars are interested in this topic and how it is suggested the use of MCDA into the GIS domain.

References

Abdul-Mawjoud AA, Jamel MG (2016) Using the analytic hierarchy process and GIS for decision making in rural highway route location. Technology 7(2):359–375

Assumma V, Bottero M, Monaco R (2019) Landscape economic attractiveness: an integrated methodology for exploring the rural landscapes in Piedmont (Italy). Land 8(7):105

Barron FH (1992) Selecting a best multiattribute alternative with partial information about attribute weights. Acta Physiol (Oxf) 80(1–3):91–103

Borgogno-Mondino E, Fabietti G, Ajmone-Marsan F (2015) Soil quality and landscape metrics as driving factors in a multi-criteria GIS procedure for peri-urban land use planning. Urban Fores Urban Green 14(4):743–750

Bottero MC, Buffoli M, Capolongo S, Cavagliato E, di Noia M, Gola M, Speranza S, Volpatti L (2015) A multidisciplinary sustainability evaluation system for operative and in-design hospitals. In: Improving sustainability during hospital design and operation. Springer, Cham, pp 31–114

Bottero MC, Comino E, Dell'Anna F, Dominici L, Rosso M (2019) Strategic assessment and economic evaluation: the case study of Yanzhou Island (China). Sustainability 11(4):1076

Bottomley PA, Doyle JR, Green RH (2000) Testing the reliability of weight elicitation methods: direct rating versus point allocation. J Mark Res 37(4):508–513

Bourne L (2005) Project relationship management and the stakeholder circle. Doctor of Project Management. Graduate School of Business

Bouyssou D (2008) Outranking methods. In: Encyclopedia of optimization. Springer, Boston, pp 2887–2893

Burkey ML, Bhadury J, Eiselt HA (2012) A location-based comparison of health care services in four U.S. states with efficiency and equity. Socio-Econ Plann Sci 46:157–163

Burrough PA, McDonnell RA (1998) Principles of geographical information systems. Oxford University Press, New York

Capolongo S, Buffoli M, di Noia M, Gola M, Rostagno M (2015a) Current scenario analysis. In: Capolongo S, Bottero MC, Buffoli M, Lettieri E (eds) Improving sustainability during hospital design and operation: a multidisciplinary evaluation tool. Springer Briefs in Applied Sciences and Technology, Cham, pp 11–22

Capolongo S, Lemaire N, Oppio A, Buffoli M, Roue Le Gall A (2016) Action planning for healthy cities: the role of multi-criteria analysis, developed in Italy and France, for assessing health performances in land-use plans and urban development projects. Epidemiol Prev 40(3–4):257–264

Capolongo S, Mauri M, Peretti G, Pollo R, Tognolo C (2015b) Facilities for territorial medicine: the experiences of Piedmont and Lombardy Regions. Technè 9:230–236

Capolongo S, Sdino L, Dell'Ovo M, Moioli R, Della Torre S (2019) How to assess urban regeneration proposals by considering conflicting values. Sustainability 11(14):3877

Danielson M, Ekenberg L (2007) Computing upper and lower bounds in interval decision trees. Eur J Oper Res 181(2):808–816

DCLG (2009) Department for communities and local government. Multi-criteria analysis—a manual. Communities and Local Government Publications, London

de Almeida AT, Almeida JA, Costa APCS, Almeida-Filho AT (2016) A new method for elicitation of criteria weights in additive models: flexible and interactive tradeoff. Eur J Oper Res 250(1):179–191

de Almeida AT, Wachowicz T (2017) Preference analysis and decision support in negotiations and group decisions. Group Decis Negot 26:649–652. https://doi.org/10.1007/s10726-017-9538-6

Dell'Ovo M, Capolongo S (2016) Architectures for health: between historical contexts and suburban areas. Tool to support location strategies. TECHNE-J Technol Archit Environ, 269–276

Dell'Ovo M, Capolongo S, Oppio A (2018a) Combining spatial analysis with MCDA for the siting of healthcare facilities. Land Use Policy 76:634–644

Dell'Ovo M, Frej EA, Oppio A, Capolongo S, Morais DC, de Almeida AT (2018b) FITradeoff method for the location of healthcare facilities based on multiple stakeholders' preferences. In: International conference on group decision and negotiation. Springer, Cham, pp 97–112

Dell'Ovo M, Frej EA, Oppio A, Capolongo S, Morais DC, de Almeida AT (2017) Multicriteria decision making for healthcare facilities location with visualization based on FITradeoff method. In: International conference on decision support system technology. Springer, Cham, pp 32–44. https://doi.org/10.1007/978-3-319-57487-5_3

Dell'Ovo M, Oppio A (2019) Bringing the value-focused thinking approach to urban development and design processes: the case of Foz do Tua area in Portugal, Valori e Valutazioni,23, SIEV, Roma, pp 91–106.

Densham PJ, Goodchild MF (1989) Spatial decision support systems: a researcha genda. In: Proceedings GIS/LIS'89, Orlando, pp 707–716

Dente B (2014) Understanding policy decisions. In: Understanding policy decisions. Springer, Cham, pp 1–127

Edwards W (1977) How to use multiattribute utility measurement for social decision making. IEEE Trans Syst Man Cybern 7(5):326–340

Edwards W, Barron FH (1994) SMARTS and SMARTER: Improved simple methods for multiattribute utility measurement. Organ Behav Hum Decis Process 60(3):306–325

Faroldi E, Fabi V, Vettori MP, Gola M, Brambilla A, Capolongo S (2019) Health tourism and thermal heritage: assessing Italian Spas with innovative multidisciplinary tools. Tourism Anal 24(3):405–419

Ferretti V (2012) Verso la valutazione integrata di scenari strategici in ambito spaziale. I modelli MC-SDSS. Celid, pp 1–174

Ferretti V, Montibeller G (2016) Key challenges and meta-choices in designing and applying multi-criteria spatial decision support systems. Decis Support Syst 84:41–52

Figueira J, Roy B (2002) Determining the weights of criteria in the ELECTRE type methods with a revised Simos' procedure. Eur J Oper Res 139(2):317–326

Gamboa G, Munda G (2007) The problem of windfarm location: a social multi-criteria evaluation framework. Energy Policy 35(3):1564–1583

Gigović L, Pamučar D, Božanić D, Ljubojević S (2017) Application of the GIS-DANP- MABAC multi-criteria model for selecting the location of wind farms: a case study of Vojvodina, Serbia. Renew Energy 103:501–521

Goodchild MF (2009) Geographic information system. In: Encyclopedia of database systems. Springer, Boston, pp 1231–1236

Greco S, Matarazzo B, Slowinski R (1999) Rough approximation of a preference relation by dominance relations. Eur J Oper Res 117(1):63–83

Greco S, Matarazzo B, Slowinski R (2001) Rough sets theory for multicriteria decision analysis. Eur J Oper Res 129(1):1–47

Greco S, Matarazzo B, Słowiński R (2009) Granular computing and data mining for ordered data: the dominance-based rough set approach. In: Encyclopedia of complexity a systems science. Springer, New York, pp 4283–4305

Greco S, Słowiński R, Stefanowski J, Żurawski M (2004) Incremental versus non-incremental rule induction for multicriteria classification. In: Transactions on rough sets II. Springer, Berlin, Heidelberg, pp 33–53

Ho HC, Knudby A, Huang W (2015) A spatial framework to map heat health risks at multiple scales. Int J Environ Res Public Health 12(12):16110–16123

Hwang CL, Yoon K (1981) Methods for multiple attribute decision making. In: Multiple attribute decision making. Springer, Berlin, Heidelberg, pp 58–191

Keeney RL (1992) Value focused thinking. Harvard University Press, Cambridge

Keeney RL, Raiffa H (1976) Decisions with multiple objectives: preferences and value trade-offs. Cambridge University Press, Cambridge

Li Y, Shen Q, Li H (2004) Design of spatial decision support systems for property professionals using map objects and excel. Autom Constr 13:565–573

Malczewski J (1999) GIS and multicriteria decision analysis. Wiley

Malczewski J, Rinner C (2015) Multicriteria decision analysis in geographic information science. Springer, New York, pp 220–228

Mat NA, Benjamin AM, Abdul-Rahman S, Wibowo A (2016) A framework for landfill site selection using geographic information systems and multi criteria decision making technique. In: AIP conference proceedings, vol 1782(1). AIP Publishing, p 040011

Mendelow AL (1981) Environmental scanning-the impact of the stakeholder concept. In: ICIS, p 20

Moghadam ST, Delmastro C, Corgnati SP, Lombardi P (2017) Urban energy planning procedure for sustainable development in the built environment: a review of available spatial approaches. J Clean Prod 165:811–827

Munda G (2004) Social multi-criteria evaluation (SMCE): methodological foundations and operational consequences. Eur J Oper Res 158(3):662–677

Oppio A, Buffoli M, Dell'Ovo M, Capolongo S (2016) Addressing decisions about new hospitals' siting: a multidimensional evaluation approach. Ann dell'Istituto superiore di sanità 52(1):78–87

Oppio A, Dell'Ovo M, Torrieri F, Miebs G, Kadziński M (2020) Understanding the drivers of Urban Development Agreements with the rough set approach and robust decision rules. Land Use Policy 96:104–678

Pawlak Z (1982) Rough sets. Int J Comput Inform Sci 11(5):341–356

Pawlak Z (1991) Imprecise categories, approximations and rough sets. In: Rough sets. Springer, Dordrecht, pp 9–32

Prasara-A J, Gheewala SH (2017) Sustainable utilization of rice husk ash from power plants: a review. J Clean Prod 167:1020–1028

Riabacke M, Danielson M, Ekenberg L (2012) State-of-the-art prescriptive criteria weight elicitation. Adv Decis Sci

Riabacke M, Danielson M, Ekenberg L, Larsson A (2009) A prescriptive approach for eliciting imprecise weight statements in an MCDA process. In: International conference on algorithmic decision theory. Springer, Berlin, Heidelberg, pp 168–179

Roy B (1974) Criteres multiples et modelisation des preferences: l'apport des relations de surclassement. Revue d'Econ Politique, 1

Roy B (1985) Multicriteria methodology for decision analysis. Kluwer Academic Publishers

Roy B (2005) Paradigms and challenges. In: Multiple criteria decision analysis: state of the art surveys. Springer, New York, pp 3–24

Saaty TL (1980) The analytic hierarchy process. Mc-Graw-Hill, New York

Stillwell WG, Seaver DA, Edwards W (1981) A comparison of weight approximation techniques in multiattribute utility decision making. Organ Behav Hum Perform 28(1):62–77

Vahidnia MH, Alesheikh AA, Alimohammadi Abbas (2009) Hospital site selection using fuzzy AHP and its derivatives. J Environ Manage 90(10):3048–3056

Varatharajan R, Manogaran G, Priyan MK, Balaş VE, Barna C (2017) Visual analysis of geospatial habitat suitability model based on inverse distance weighting with paired comparison analysis. Multimedia Tools Appl, 1–21

Von Winterfeldt D, Edwards W (1986) Decision analysis and behavioral research. Cambridge University Press, Cambridge

Walley P (1991) Statistical reasoning with imprecise probabilities. Chapman and Hall, London

Yang N, Liao X, Huang WW (2014) Decision support for preference elicitation in multi-attribute electronic procurement auctions through an agent-based intermediary. Decis Support Syst 57:127–138

Yang RJ (2014) An investigation of stakeholder analysis in urban development projects: empirical or rationalistic perspectives. Int J Project Manage 32(5):838–849

Chapter 3
Approaching the Location of Healthcare Facilities: How to Model the Decision Problem

Abstract The management of health policies is characterised by multi-level hierarchies and actors. Given the presence of several and sometimes conflictual needs and expectations elicited by stakeholders involved and the complexity of the decision problem concerning the location of healthcare facilities, a multi-methodological evaluation framework is proposed within this contribution. The process is based on the four traditional stages of the Multi-Criteria Decision Analysis (MCDA), recognised as the most critical in the context of the location of healthcare facilities, namely stakeholder analysis, criteria definition, weights assignments and aggregation procedure. Since several methodologies have to be combined, the proposed framework is conceived as an iterative and flexible approach where each phase can be reviewed when necessary. The methodology proposed is tested on a case study located in the Municipality of Milan, Italy.

3.1 Introduction

In Italy, regional bodies are in charge to manage and take decisions about health policies and planning actions, but usually conflicting interests with local authorities arise. In fact, one of the main reason why the location of healthcare facilities can be considered as an ill-structured problem (ISP) (see Chap. 1) is due to the presence of multiple and often conflictual stakeholders, due to a multi-level hierarchies and actors. Crivellini and Galli (2011) have given an overview about the healthcare system in developed countries and have provided a scheme about the institutional actors in charge to manage this topic and their relations. In detail the framework is divided into four levels and five elements:

- Central level: central bodies of the state;
- Intermediate level: intermediate bodies and actors; economic-financial subjects for the sale and purchase of health services;
- Provider level: facilities and actors that provide health services;
- Users level: citizens.

More in detail:

- Citizens are the key element since they are those who finance the health system;
- Facilities and actors provide the medical and health assistance;
- Economic-financial subjects: they could be both private and public and are aimed at buying and selling health services (e.g. insurances, etc.);
- Intermediate bodies and actors: according to the role represented in each corresponding country, could have different commitments and purposes;
- Central bodies of the state: they strongly determine the healthcare system adopted by the country.

The complexity of taking a decision about healthcare management and planning has been underlined within the institutional context, by Gola et al. (2018). Moreover, with respect to the Italian context, the National Health Service—Servizio Sanitario Nazionale (SSN)—established by the law n. 833 in 1978, is organised and managed mainly by State and Regions each one with special commitments. In detail the State is in charge to guarantee the citizens' right to health through an efficient system of healthcare, while Regions have the direct responsibility to achieve this objective by defining special criteria to fund local health companies and hospitals. Sometimes the roles can interfere each other since interests and responsibilities can overlap. Moreover, the SSN involves other bodies and authorities of different institutional levels, thus increasing the governance complexity because of the interdependencies between actors, resources and activities when achieving common goal by combined effects.

The annual report developed by SDA Bocconi, School of Management (Rapporto OASI 2017, Osservatorio sulle Aziende e sul Sistema sanitario Italiano) highlights that the Regions in addition to their administrative role have the task to: (a) identify health companies as a part of the regional and public health group; (b) delegate the economic management of health companies; (c) structure the regional administrative framework and (d) organise the operations of the companies that are part of the group. Moreover, sometimes Regions establish Health Regional Agencies (Agenzie Sanitarie Regionali) to further supervise the health territorial organisation and to monitor its performances.

The Italian institutional framework further explains the reason why the health management and the decisions regarding the location of healthcare facilities has been assigned to ISP.

The aim of the current section is to test some of the methodologies presented in the second chapter in order to understand their suitability to set and solve the decision problem concerning the location of healthcare facilities and to propose a multi-methodological framework to support the Decision-Maker (DM).

The paper is divided in five sections. After a brief introduction about the Italian hierarchical institutional framework, Sect. 3.2 will present the methodology proposed, subsequently tested in Sect. 3.4 on a case study located in the Municipality of Milan, Italy (presented in Sect. 3.3). Final conclusions (Sect. 3.5) will discuss the main achievements and criticalities detected by the application.

3.2 Multi-methodological Evaluation Framework

Given the already mentioned complexity of the decision problem, it is clear that it is not possible to solve it in a linear way, but according to an iterative approach.

Within this contribution, methodologies previously described (see Chap. 2) are going to be applied to a case study according to four stages of the Multi-Criteria Decision Analysis (MCDA). Since several methodologies have to be combined, the proposed framework is multi-methodological (Bottero et al. 2015). It is conceived as an iterative and flexible approach where each phase can be reviewed when necessary. Figure 3.1 shows how different methodologies have been combined and different phases interact (Bottero et al. 2020).

The proposed process is therefore based on the four traditional stages of the MCDA, recognised as the most critical in the context of the location of healthcare facilities.

The first phase considers the analysis of the actors involved in the decision problem. More in detail they have been classified according to Dente (2014) into five categories and their scale of action has been highlighted. In fact, here the complexity of the problem is more explicit: having actors working at different scales increases the number of points of views to be analysed. Once the actors have been identified, it is also important to understand their level of power and interest in order to assess which of those identified are the most influential and their internal relationships.

For what concerns the second phase, through the analysis of the tools and literature review and subsequently according to the elicitation of the opinions of the actors involved, it has been possible to define an appropriate set of criteria and indicators. The MCDA allows to break down the problem into hierarchical levels to better manage its complexity, as suggested by the literature.

During the third phase different weights elicitation methods have been applied to test how they perform for this specific decision context. The first method considers the online administration of a questionnaire through which from a list of criteria, it has been required to select the most important and to assign them a score. Then the results

Fig. 3.1 Methodological flowchart

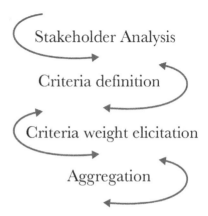

have been grouped considering the classification obtained by the power/interest matrix and weights have been assigned. The second methodology is the FITradeoff method (de Almeida et al. 2016) developed by a research group of the Universidade Federal de Pernambuco in Recife (Brazil) which is mainly divided into two phases. In the first is asked to the DM to order the criteria from the most important to the least important. To facilitate the operation, the comparison in pairs, instead of the rank order method, can be selected. Once ordered, the software incorporates the preferences. Since alternatives and the related performances are processed by the method, once the first phase is finished, a solution can already be found, otherwise the second phase is presented, in which two scenarios of possible consequences with different performances are compared and the DM is called to decide which of the two prefers. Between one question and another the software processes the preferences until there is only one solution left. The method has been tested with an actor representing each class of the power/interest matrix. The third methodology is the Point allocation (Bottomley et al. 2000) which has been administered to a panel of experts with expertise in decision making and considered competent and able to give objective and not personal judgments.

In the fourth phase three aggregation procedures have been tested and pros and cons explored, such as the possibility of evaluating both the aggregate and the partial results in addition to the critical points such as the inconsistency detected or the excessive subjectivity of some sub-steps. In fact, for two methods, the Analytic Hierarchy Process (AHP) (Saaty 1980) and the Weighted Sum (Triantaphyllou 2000; Fishburn 1967), a sensitivity analysis has been performed while for the FITradeoff method (de Almeida et al. 2016), through the visualization of the Weight Space, it is possible to understand for which values the result remains stable.

3.3   Case Study: La Città Della Salute, Milan, Italy

In order to test the validity of the methodologies previously defined, a case study concerning the location of a healthcare facility in the Municipality of Milan, Italy, has been selected. The pilot study considered the location of "La Città della Salute" which aimed at relocating two existing hospitals—Istituto Neurologico Carlo Besta and Istituto dei Tumori—in order to create a unique centre of research and cancer treatments (Fig. 3.2). The final purpose of this project was to combine innovativeness as well as clinical and scientific excellence in a public service (Dell'Ovo et al. 2017, 2018a, b; Oppio et al. 2016).

The study started with a spatial analysis within the Municipality of Milan to understand the distribution of existing healthcare facilities (Dell'Ovo and Capolongo 2016) and important hints have been detected:

- there are around thirty public hospitals with different specializations;
- the majority has been realized between the 50s and the 90s;

Fig. 3.2 Location of the case study

- for outdated hospitals the structure is not adequate anymore to host health functions;
- they are well distributed in the consolidated city and in suburban areas (54% vs. 46%).

After this preliminary analysis the reasons behind the final decision to locate "La Città della Salute" in the site of Sesto San Giovanni has been deeply investigated. Despite the real choice has already been made, the decision underwent a long and not transparent process, guided moreover by political and economic instances, rather than by a comprehensive analysis of the other potential five areas. The aim of this study is not to discuss the choice taken, but to show how it should be developed a process to support DMs in solving a complex location problem. The investigation demonstrates how the support of a preliminary feasibility analysis and the use of robust methodologies could improve the transparency and the efficiency of decision problems. A deep description of the case study and its relative timeline has been already developed by Dell'Ovo et al. (2018a, b), where the main important phases of the process are detailed.

3.4 Application of the Methodological Framework

Within this section the four phases of the multi-methodological framework will be applied to solve the problem concerning "La Città della Salute" in the Municipality of Milan, Italy. For each stage it will be tested the methodologies described in the second Chapter and the main achievements will be later discussed.

3.4.1 Stakeholder Analysis

The stakeholders involved in the location of healthcare facilities and in particular of "La Città della Salute" have been investigated according to the classification of Dente (2014) into five distinct categories. The analysis of the decision problem has been carried out so as to identify for each category the corresponding actor in the real world. Actors that could affect or be affected by the project under investigation are presented in Table 3.1.

As suggested by Dente (2014) and explained in the second chapter "we could run into 'methodological constitutionalism' in case only actors that are legally responsible for implementing a project are considered". Thus, the whole set of stakeholders have been considered, also NGO and NPO aimed to "represent who is not able to defend themselves" and citizens, common people with their specific interests and with the purpose to easily access to care. Moreover, in Table 3.1 it is possible to appreciate the scale of action of stakeholders involved in the decision problem. This further analysis allows to understand the complexity of the problem, i.e. the plurality of points of view. If actors are of the same category and work at the same level, even if owners of different objectives, they will tend to evaluate the problem in a homogenous way, eliciting the same set of criteria. On the other hand, a plurality and variety of categories and levels will present a more varied situation. In this specific case there is a high level of complexity because all the categories and the levels are filled.

A further analysis concerns the investigation of the level of interest and power (Mendelow 1981) of the five categories previously defined. In this phase, the actors are further classified into four classes according to their position in the power/interest matrix of, as can be seen in Fig. 3.3.

Table 3.1 Identification of stakeholders and scale of action

Category	Actor	Scale of action
1. Political actors	Health and urban councillor	Regional
2. Bureaucratic actors	Health and urban general manager	Regional
3. Actors with special interests	Local health unit director	Local; province
4. Actors with general interests	Common people	Local; province; national
5. Actors with general interests	NPO, NGO	National; international
6. Experts	Architects, urban planners, technology expert, transport policy expert, urban economist, project appraisal expert, environmental hygiene expert, management engineer, environmental practice expert	Local; province; national; international

LEVEL OF INTEREST

		low	high
POWER	low	Stakeholder: **Minimal effort** Nonprofit Organizations; Non-governmental Organizations	Stakeholder: **Keep informed** Common people
	high	Stakeholder: **Keep Satisfied** Health and Urban General Manager, Local Health Unit Director, Experts	Stakeholder: **Key players** Health and Urban Councilor

Fig. 3.3 Matrix power/interest related to the location of healthcare facilities

Figure 3.3 shows the hierarchy of power with respect to the decision problem under investigation. The key players are represented by political actors since the have the power to take the decision and also a high interest given by their institutional role. The power of the bureaucratic actors and experts is also high, they have specific competences and knowledge and they can strongly influence the final choice, while they could not have a high interest in relation to this case study. On the contrary NPOs/NGOs and common people own a high interest, as the location of hospitals can affect their daily life, but the power of influencing the final decision is low.

Instead, Fig. 3.4 represents the Circle methodology (Bourne 2005) and it is immediately clear that stakeholders with the power to kill the project are the political and the bureaucratic one. Those one with special interest are close to the project but with limited individual influence. Common people have a significant power as a group but not as individuals. NGOs and NPOs even if located distant from the project can have a particular influence on it and experts have a limited influence but with their competences can kill the project.

The main results achieved in this phase concerns the comprehension of all possible clients and users, the strength of their influence in solving the decision problem and their internal relationship. Once framed this basic concept it is possible to have a direct interaction with them in order to elicit their values and understand their wills.

From this analysis it is possible to perceive who has to be mostly satisfied and from whom they are influenced.

Each problem can be represented by different actors but this general framework can help to identifying the specific set since it is able to clarify roles and competences. With this kind of analysis every DMs could be able to understand immediately with who they should talk and who should satisfy in taking a decision since here everyone affected by the problem is described. Maybe case by case some roles could change

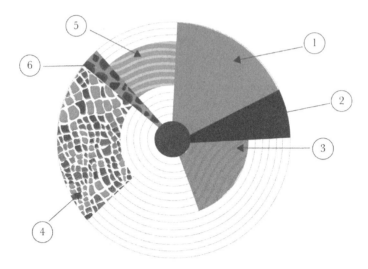

Fig. 3.4 Stakeholder Circle methodology referred to the hospital's location (legend: 1: Political actors; 2: Bureaucratic actors; 3: Actors with special interests; 4: Actors with general interests—common people; 5: Actors with general interests—NGO, NPO; 6: Experts)

according to the client, the context, the political structure, but the categories and the purpose they should achieve is more or less fixed.

3.4.2 Criteria Definition

This phase recalls the first chapter where a review about the literature and existing evaluation tools have been developed in order to understand the methodologies applied by other scholars but also to list of criteria already used. In fact, here a suitable set of criteria according to the aim of locating a healthcare facility is going to be defined in order to solve the decision problem. Some of the methodologies already described here are merged with the purpose to use a unique set. Figure 3.5 shows the flowchart used to define a suitable set of criteria. How it going to be discussed below, the process took into consideration both most frequent and appropriate criteria present already in the existing evaluation tools selected and in the literature analysed. Moreover, a further step considered the opinion of stakeholders about the decision problem, eliciting what they aspire to achieve by the location of healthcare facilities.

As it is possible to appreciate from the flowchart (Fig. 3.5) criteria defined by the evaluation tools and the literature review (Dell'Ovo et al. 2017, 2018a, b; Dell'Ovo and Capolongo 2016; Oppio et al. 2016) are going to be compared in order to check the presence of similarities or overlapping and then, to further validate the set defined, the opinion of stakeholders has been considered.

Fig. 3.5 Flowchart for the criteria definition

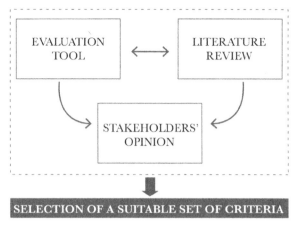

The process to frame a suitable set of criteria to solve the decision problem underwent a deep analysis of the State of the Art of existing evaluation tools that are focused on assessing the energy performance of hospitals. The analysis was supported by a review of the literature aimed to locate healthcare facilities.

A comparative analysis has been performed to highlight and define the most used criteria by evaluation tools (e.g. LEED Healthcare, BREEAM Healthcare, Metaprogetto DM 12/12/00, etc.) and cited by the analysed literature. Below is presented a graphic visualization of the results obtained (Fig. 3.6). In particular it is possible to notice how the distinction between tools and literature review has been considered by underlying direct connections with the red line and with the blue the indirect ones. Direct connections mean when in both the analysis the same criterion with the same description has been found out, while for indirect when a correlation is present, for example one factor is the consequence of the presence of another one.

This result allows to understand how there are some essential characteristics to take into consideration while others could be considered redundant since already included in the description of other factors.

Moreover, according to the stakeholders previously identified, their position regarding the location of hospitals (Table 3.2) have been elicited in order to understand their interpretation of the problem and how to satisfy their instances (Gamboa and Munda 2007).

Table 3.2 highlights the values expressed by the actors involved, why they care about the decision and which aspects they would like to maximize. As an example the position of common people can be cited. They "prefer places easily accessible", i.e. they will promote the accessibility as criterion to consider in the evaluation of suitability of a site.

How it is possible to appreciate from the previous analysis different methodologies have been combined in order to result in a unique set of criteria. Understanding the position of actors regarding the location of healthcare facilities allows to elicit both

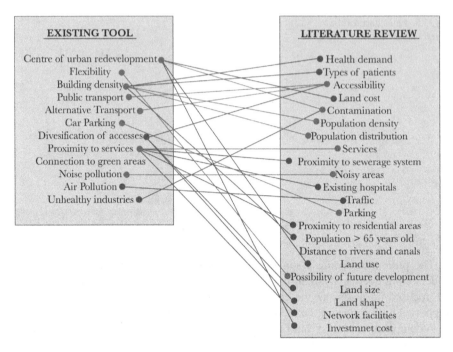

Fig. 3.6 Comparative analysis between literature review and existing evaluation tools

their values-objectives and their expectations as suggested by Gamboa and Munda (2007) and Keeney (1992).

A set of sixteen criteria has been defined by taking into consideration four different dimensions of the problem—Functional; Locational; Environmental and Economic—then related to appropriate criteria (Tables 3.3 and 3.4). According to the definition of criteria within the MCDA (Roy 1985, 2005), it is easier to decompose a problem in hierarchies and smaller elements to facilitate its comprehension. This is the reason why four macro-areas have been found out in order to collect other information. Table 3.3 shows both the decision tree and the scientific or operational source for each variable. As it is possible to appreciate in the third column, the tools have been specified by their acronym while for the scientific literature authors (see Chap. 1) has been cited. This step wants to remark the importance of studying the State of the Art and how it has been useful to frame the current set of criteria proposed.

Moreover, in Table 3.4 indicators have been added to the framework, in order to define how criteria are going to be evaluated. In this phase it is fundamental to be more precise as possible both in case of qualitative and quantitative measurements, for avoiding misunderstandings and carrying on a clear and transparent procedure. Since this framework could be used by other actors (non-expert with this kind of methodology), then they should feel confident and using it in an appropriate way.

Table 3.2 Position of actors regarding the hospitals' location

Actor	Position regarding hospitals' location
Health and urban councillor	Representing citizens, health and urban councillors prefer to locate hospital in place in accordance with the needs of the population
Health and urban general manager	Health and urban general manager will choose a place on the basis of their interpretation of the law, on the observance of legally predetermined procedures and on the respect of the roles they defined
Local health unit director	The hospital location could affect the competitiveness of it and then its management. Local health unit directors want to locate hospitals in strategic places
Common people	Common people prefer places easily accessible and near services
NPO, NGO	NPO and NGO prefers sites to locate hospitals in accordance with sustainable principles and able to respect civil rights
Architects, urban planners, technology expert, transport policy expert, urban economist, project appraisal expert, environmental hygiene expert, management engineer, environmental practice expert	Experts according to their specific competences will choose a site able to respect specific requirements (e.g. urban and environmental constraints)

Once the set of criteria has been defined, it is important to evaluate if the framed decision tree is able to satisfy stakeholders' expectations, and in particular who and how. Table 3.5 allows to visualize which actor "promotes" each criterion and which expectations they aim to achieve by that promotion. In particular what emerges from the table is the reason why that criterion is important, why it is worth to be included in the evaluation framework, what it is possible to reach by pursuing that objective. Not every criterion is supported by all the categories of stakeholders. According to their role and knowledge they behave differently. For example, actors that own economic resources have interests in promoting the economic dimension. Only the accessibility, belonging to the Locational dimension is supported by the five categories of actors because everyone has interests in having a healthcare facility easily accessible by all the typologies of different kind of mobility.

3.4.3 Criteria Weight Elicitation

Given the set of criteria, the criteria weight elicitation has been developed by considering the influence of each criterion in achieving the final aim of the location of healthcare facilities.

Table 3.3 Set of criteria and from whom they are suggested

Dimension	Criteria	Suggested by
Functional quality	Building density	Leed (2011), Breeam (2010), (Vahidnia et al. 2009; Wu et al. 2007; Lee and Moon 2014; Zhang et al. 2015; Noon and Hankins 2001; Murad 2005; Burkey 2012; Chiu and Tsai 2013; Beheshtifar and Alimoahmmadi 2015; Faruque et al. Faruque et al. 2012; Soltani and Marandi 2011; Kim et al. 2015)
	Health demand	(Murad 2007; Burkey et al. 2012; Wu et al. 2007; Lee and Moon 2014; Murad 2005; Chiu and Tsai 2013; Beheshtifar and Alimoahmmadi 2015; Du and Sun 2015; Kim et al. 2015)
	Reuse of built-up areas	Leed (2011), Breeam (2010), Metaprogetto, Progetto Itaca, Casbee
	Potential of the area to become an attractive pole	Leed (2011), Breeam (2010), Metaprogetto, Progetto Itaca, Casbee
Location quality	Accessibility (private, public, parking)	Leed (2011), Breeam (2010), Metaprogetto, Progetto Itaca, Casbee, (Murad 2007; Burkey et al. 2012; Vahidnia et al. 2009; Wu et al. 2007; Lee and Moon 2014; Abdullahi et al. 2014; Wu et al. 2012; Murad 2005; Burkey 2012; Chiu and Tsai 2013; Beheshtifar and Alimoahmmadi 2015; Daskin and Dean 2005; Soltani and Marandi 2011; Du and Sun 2015; Kim et al. 2015)
	Existing hospitals	(Vahidnia et al. 2009; Lee and Moon 2014; Zhang et al. 2015; Abdullahi et al. 2014; Shariff et al. 2012; Murad 2005; Burkey 2012; Chiu and Tsai 2013; Daskin and Dean 2005; Soltani and Marandi 2011; Du and Sun 2015; Kim et al. 2015)
	Services	(Noon and Hankins 2001; Kim et al. 2015)
	Sewerage system	(Abdullahi et al. 2014; Kim et al. 2015)
Environmental quality	Connection to green areas	Leed (2011), Breeam (2010), Metaprogetto, Progetto Itaca, Casbee
	Presence of rivers and canals	(Abdullahi et al. 2014)

(continued)

Table 3.3 (continued)

Dimension	Criteria	Suggested by
	Air and noise pollution	Breeam (2010), Casbee, (Zhang et al. 2015; Abdullahi et al. 2014; Soltani and Marandi 2011)
	Land contamination	(Vahidnia et al. 2009; Zhang et al. 2015; Abdullahi et al. 2014; Beheshtifar and Alimoahmmadi 2015)
Economic aspects	Land size and shape	(Abdullahi et al. 2014; Soltani and Marandi 2011)
	Land ownership	(Wu et al. 2007)
	Land cost	(Vahidnia et al. 2009; Wu et al. 2007; Chiu and Tsai 2013; Du and Sun 2015)
	Land use	(Wu et al. 2007; Soltani and Marandi 2011)

For this specific phase different methodologies have been applied. After the applications, since specific suggestions have been received by the interaction with stakeholders interviewed and given the iterative characteristic of the multi-methodological evaluation framework, the set of criteria has been reviewed.

Stakeholders have been analysed in two different steps.

The first time to understand their needs and expectations in relation to the location of healthcare facilities in order to define criteria. The second time to weight the criteria. Starting from the idea to test different methods in order to understand which one could be the most appropriate according to the decision context and the stakeholders involved, different procedures have been applied to the specific case study of the location of "La Città della Salute" in order to highlight pros and cons for each of them.

In particular two methodologies are going to be explained, one related to the administration of an online questionnaire and the other one by applying an innovative procedure called FITradeoff method (de Almeida et al. 2016).

Online questionnaire

In order to understand stakeholders' preferences, it has been sent an online questionnaire by asking to distribute 10 points among the criteria they consider the most important (Rebecchi et al. 2016). The results have been aggregated according to the votes obtained by each criterion to create an unique rank. This method has been considered the most appropriate given the high number of criteria to evaluate and also to discard those ones ranked at the last places.

The questionnaire has been structured into two phases. In the first part of the questionnaire the decision problem has been explained in order to provide a general overview. In the second part (Fig. 3.7) stakeholders were asked to assign a maximum of 10 point to the criteria they consider fundamental for the location of healthcare

Table 3.4 Set of criteria

Dimension	Criteria	Performance
Functional quality	Building density	Number of people living in and near the area
	Health demand	Percentage of people above 65 years of age
	Reuse of built-up areas	Promotes the use of sites already exploited
	Potential of the area to become an attractive pole	Encourages the development of peripheral and degraded sites
Locational quality	Accessibility	Public, private and soft mobility and number of parking lots
	Existing hospitals	The presence could be an advantage or disadvantage depending on the project
	Services	Number of specific facilities present in a radius of 800 m from the site
	Sewerage system	Presence of this infrastructure
Environmental quality	Connection to green areas	Correspondence to specific characteristics (e.g. the possibility to reach parks and garden in a short time)
	Presence of rivers and canals	To avoid the choice of sites with hydraulic and hydrological instability
	Air and noise pollution	Concentration of specific pollutants and the dB level detected by surveys on site
	Land contamination	High-medium-lo, in line with the site being suitable to host a hospital
Economic aspects	Land size and shape	Ratio between the dimension of the site and that one of the new hospital
	Land ownership	Percentage of public and private areas
	Land cost	Price (€/sqm)
	Land use	High-medium-low in relation to the tendency for the site to host new healthcare facilities in accordance with the type of land cover

Table 3.5 Stakeholders' needs and expectations

Dimension	Criteria	Actors	Needs and expectations
Functional quality	Building density	Expert	– To promote sustainable development – To locate healthcare facilities where are needed – To preserve farmland and sensitive areas
	Health demand (related to types of patients and population age)	Bureaucratic, expert	– To locate healthcare facilities where are needed – To take into consideration people disease
	Reuse of built-up areas	General interest, expert	– To promote sustainable development – To preserve farmland and sensitive areas – To reactivate the economic dynamics
	Potential of the area to become an attractive pole	General interest	– To attract people in the neighbourhood – To reactivate the economic dynamics
Location quality	Accessibility (private, public, parking)	Political, bureaucratic, special and general interest and expert	– To allow any type of users to reach the area – To promote public and soft mobility – To reduce pollution and traffic
	Existing hospitals	Bureaucratic, expert	– To avoid conflicts and competitiveness – To take into consideration the catchment area
	Services	General interest	– To avoid the feeling of isolation – To attract people in the neighbourhood
	Sewerage system	Political	– To prevent hygiene problem and infections

(continued)

Table 3.5 (continued)

Dimension	Criteria	Actors	Needs and expectations
Environmental quality	Connection to green areas	General interest, expert	– To reduce stress of hospitals' users – To promote sustainable development – To improve quality life of patients
	Presence of rivers and canals	Political, expert	– To prevent hydraulic and hydrogeological instability
	Air and noise pollution	General interest	– To protect human health
	Land contamination	General interest	– To prevent additional costs – To promote sustainable development
Economic aspects	Land size and shape	Political, special interest	– To minimize costs
	Land ownership	Bureaucratic	– To prevent additional costs – To consider areas' fragmentation
	Land cost	Special interest, expert	– To rationalize resources – To quantify the value of the area
	Land use	General interest, expert	– To consider areas' intended use – To minimize costs

facilities. The questionnaire has been sent to around 100 stakeholders belonging to the five categories and it has been filled only from 44 actors. Even if through the use of email, it is possible to reach a large audience, it could be risky since very often requests to fill questionnaire are suddenly deleted.

Answers have been analysed according to the categories of stakeholders interviewed (Fig. 3.8) and then merged together in order to get to a unique value. Since also stakeholders show different power to affect decisions, it has been assigned to them a different degree of importance by considering the power and interest matrix.

In particular, to aggregate the results, below the weights assigned to each stakeholder in relation to the power and interest matrix are shown (Fig. 3.3):

- Key player: 40%;
- Keep satisfied: 25%;
- Keep informed: 25%;

Fig. 3.7 Second part of the online questionnaire

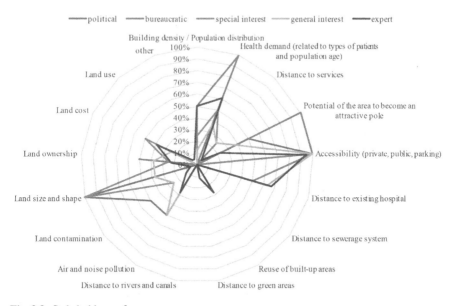

Fig. 3.8 Stakeholder preferences

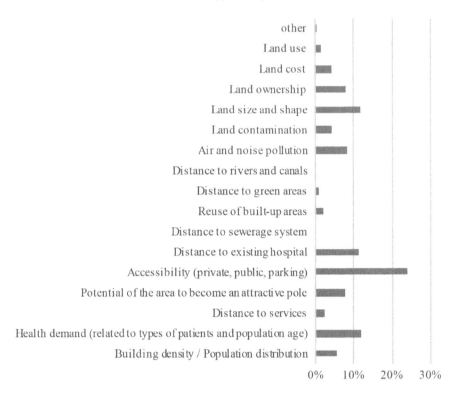

Fig. 3.9 Weights assignment

- Minimal effort: 10%.

  Figure 3.9 shows the final rank obtained.

The FITradreoff Method

FITradeoff (de Almeida et al. 2016) is a Flexible and Interactive Tradeoff elicitation procedure for multicriteria additive models that uses partial information about
DMs preferences to determine the most preferred alternative among a defined set
by considering an additive model in Multi-Attribute Value Theory scope—MAVT
(Keeney and Raiffa 1976). Compared to the traditional trade-off methods, it aims at
minimizing the inconsistency by reducing the DM's cognitive effort. In fact, during
the interaction with the DM, it is asked to answer only to questions about strict
preference relations.

This methodology has been applied two times and it is possible to appreciate the
results in Dell'Ovo et al. (2017, 2018b). In the first paper the methodology has been
applied with one actor, while in the most recent one with one representative of each
category resulted from the matrix of power and interest.

As it has been described in the second section, the methodology is divided into
steps. In the first one the DM is asked to rank the set of criteria according to their level

of importance. The answers are incorporated in the linear programming problems (LPP) model and at this point there are two possibilities. Since the methods requires the identification of a set of alternatives evaluated according to their performances, it could happen that a unique solution is found after the ranking phase, considering the first weight space and then ending the process. Otherwise, the DM can accept the partial results and compare the remaining alternatives, to decide whether he wants to continue the process or stop it. If the DM decides to continue, questions about preference relations between consequences are asked. The answers are incorporated in the LPP model as constraint and considered in the weight space. The process goes on, until a unique solution is found or DM is not willing to proceed.

The two methodologies applied for the criteria weight elicitation have been compared according to their own characteristics (Table 3.6) and by considering strengths and weaknesses (Table 3.7).

There are not correct and wrong methods but of course, according to the decision problem to solve, one can be more suitable than another one. Anyway here it is not possible to choose one procedure instead of another one since it has been recognized an underlying problem during the definition of the set of criteria. In fact, both the methodologies highlighted the high number of criteria, some redundancy and at the same time some correlations. Given the flexibility of the proposed process, based on the four traditional stages of the MCDA, its phases can be reviewed any time, in fact, in this case it has been necessary to come back to the Criteria definition phase and review the set of criteria already defined. The flowchart in Fig. 3.1 has been reviewed and the criteria modified.

According to suggestions and feedback received, Table 3.8 shows the final set of criteria with the assigned weights.

In detail, within the Functional dimension, Centre of urban redevelopment aggregates information regarding two criteria present in the previous set: Reuse of Built-up areas and Areas with potential to become attractive (Capolongo et al. 2019); Flexibility (Buffoli et al. 2012) can be compared to Land size and shape; Building density

Table 3.6 Comparative analysis of weight elicitation methods

	Method 1: questionnaire	Method 2: FITradeoff
Weight elicitation	Point allocation and rank-order method	Pairwise comparison or rank-order + tradeoff
No. stakeholder	44	1 per time (tot. 4)
No. categories of stakeholder	5	1 per time (tot. 4)
Technique used to determine the weights	Questionnaires	Interactive interview
Interaction with stakeholder	Low	High
Comprehension of the decision problem	Low	High
Time (to answer)	5 min	Depend
Effort for the stakeholder	Low	Medium

Table 3.7 Weight elicitation methods strengths and weaknesses

	Method 1: questionnaire	Method 2: FITradeoff
Weight elicitation method	Point allocation and rank-order method	Pairwise comparison or rank-order + tradeoff
Characteristics	The questionnaire allows to reach many stakeholders but it was not possible to explain in deep the decision problem and the meaning of different criteria. Answers were divided according to the categories of stakeholders previously defined. It is useful to check the robustness of the results, but at the same time the inconsistence could be of medium level according to the degree of comprehension of criteria	The interactive interview allows to explain in deep the decision problem and the meaning of each criterion. In particular the DM has been guided across all the process without the risk of misunderstanding some steps. The effort required has been judged of medium level. The high level of interaction allows the DM to stop the procedure any time when he feels satisfied from the result. Reducing DM's cognitive effort is a way to minimize inconsistencies

Table 3.8 Criteria weights assignment

Dimensions	Weights (%)	Criteria	Weights (%)
Functional quality	23	Centre of urban redevelopment	9
		Flexibility	64
		Building density	27
Location quality	34	Accessibility	44
		Services	9
		Green area	4
		Network infrastructures	43
Environmental quality	25	Noise pollution	24
		Air pollution	63
		Unhealthy industries	13
Economic aspects	18	Value of the area	14
		Land ownership	43
		Land suitability	43

was already present but here combines also data about Health demand. Criteria of the Location dimension are almost the same, except for Green areas that have been moved here from the Environmental dimension and took the place of Existing hospital. Within the Environmental dimension, Air and noise pollution have been split to create two different criteria given their importance for hospital location and another important element added, given its impact on people health is the presence of Unhealthy industries (Gola et al. 2019). Economic aspects concern variables that can affect the cost of preliminary operations: Value of the area, Land ownership and Land suitability that is similar to Land contamination present in the previous set. As it is possible to understand some criteria have been aggregated since considered as redundant, others better described, while other two—Existing hospital and Presence of rivers and canals—removed. They are not anymore present because introduced as a prerequisite in the analysis in order to exclude a priori the areas not suitable.

The criteria weights' elicitation has been carried out by considering a panel of experts through the point allocation method (Bottomley et al. 2000). Experts were asked to distribute a total of 100 points among the criteria according to their level of importance in affecting the location problem.

It has been selected this methodology because considered as appropriate in the context of the location of healthcare facilities given the possibility to elicit the opinion of different typologies of "experts" with no specific and personal interest on the decision problem under investigation. Moreover, the point allocation resulted both of easy comprehension for the DM and easy to explain for the analyst. In particular to weight the criteria one expert for each dimension has been interviewed:

- Functional quality: expert in hospital design and public health;
- Location quality: expert in urban planning;
- Environmental quality: expert in outdoor and environmental quality hygiene;
- Economic aspect: expert in real-estate market and feasibility studies.

For what concerns the four dimensions, a total of ten actors have been interviewed, the previous four and other six belonging to the other categories identified in Sect. 4.1. The group of identified actors has been able to represent all the issues defined and to combine the outputs of stakeholder analysis with the criteria definition.

In the set of criteria now identified, there are no more redundancies and even if it may seem that correlations are present, through the testing phase, it has been possible to verify that they are not closely correlated.

Even for the Point allocation, the characteristics detected for the other two methodologies can be explained, in detail:

- N° Stakeholders: 10;
- N° of Categories of Stakeholder: 3
- Technique used to determine the weight: Panel of expert;
- Interaction with Stakeholder: medium;
- Comprehension of the decision problem: medium;
- Time (to answer): 15 min;
- Effort for the stakeholder: medium;
- Characteristics: The panel of expert allows to explain the decision problem before starting to answer but without going in deep on the meaning of each criterion. Stakeholders understood the methodology and inconsistence have not been detected.

Table 3.9 presents a brief description of the evaluation framework.

Table 3.9 Description of the evaluation framework

Dimensions	Criteria	Description
Functional quality	Centre of urban redevelopment	Abandoned and built-up areas are preferred in order to promote urban regeneration
	Flexibility	It considers the possibility of future expansion of the hospital
	Building density	Urban areas are preferred
Location quality	Accessibility	It considers the private, public and soft mobility, the presence of parking lot nearby the area and the diversification of access to reach the site
	Services	To promote the inclusion of the hospital in people daily life. The performance considers the presence of different typologies of services in a radius of 800 m from the site
	Green area	To promote the integration of the hospital with the natural environment
	Network infrastructures	It evaluates the provision and adequacy of sewerage system nearby the area
Environmental quality	Noise pollution	The performance is calculated considering the limit thresholds given by local regulations
	Air pollution	Three different pollutants have been considered PM10, O3 and NO_2
	Unhealthy industries	To avoid their presence nearby the hospitals
Economic aspect	Value of the area	The monetary value of the site is calculated
	Land ownership	It is considered the percentage of public and private ownership; the public one is favoured
	Land suitability	It evaluates if preliminary works are necessary to use the area

3.4.4 Aggregation Procedures

Given the above described framework, it is possible to start the alternatives' performances evaluation and to aggregate scores and weights.

For what concerns the location of "La Città della Salute", a total of six alternatives have been screened before selecting Sesto San Giovanni as the final one, (Fig. 3.2) and within this section they will be evaluated and compared. In detail, once all the alternatives have been assessed considering specific indicators (Oppio et al. 2016) it has been possible to proceed with the analysis. Different aggregation procedures have been applied and tested, in particular: the AHP developed by Saaty (1980), the weighted sum and the FITradeoff method. Since the aim of this research is to test methodologies, in order to understand the most appropriate one, it will not be explained all the application but main phases and results.

Analytic Hierarchy Process (AHP)
The AHP allows to obtain as a result a ranking of alternative choices on the basis of pairwise comparison between the constituents of the decision problem. The model is able to process many and mixed information, both qualitative and quantitative, by standardizing the scores obtained by alternatives. The pairwise comparison, based on the 9 points fundamental Saaty scale, can be the result of actual measurement or could reflect the relative strengths of preferences (Saaty 1987). Also the weights assignment is based on pairwise comparison of all the elements for all the hierarchical levels. Finally, it is possible to aggregate the intensity obtained by each alternative against each criterion and the weights previously defined in order to obtain both partial and total rank. In general, the AHP is conceived as a framework for developing "deductive and inductive thinking" by allowing numerical tradeoff to achieve a final conclusion (Saaty 1987).

Weighted Sum
In the Weighted Sum method, the value functions are summed up with varying weights and then the sum is optimized. In fact, it combines all the multi-objectives function into a composite objective function (Yang 2014). This aggregation procedure is able to generate a ranking of the alternatives based on the scores obtained by each alternative against one criterion and their assigned weights. This is one of the most common and frequently used aggregation method.

Since criteria have different units of measurement, one of the first step is the standardization. Different softwares have been developed to facilitate these operations and for this specific case study the Definite Software (Janssen et al. 2000) has been applied. It allows to follow all the passages required by the MCDA. Moreover, it is possible to perform a sensitivity analysis able to check the internal robustness of the results.

The FITradeoff method
The FITradeoff method (de Almeida et al. 2016) has been already explained in the previous sections but without presenting the results.

After the application of the two phases (criteria ranking and question-answering step) a solution can be find. In particular, by considering the second step every time the DM answers compering two consequences, the LPP incorporate that preference as constraints. With these constraints, a weight space is obtained, which is updated with DM responses in order to reduce the subset of potentially optimal alternatives. The interaction goes on until a unique solution is not found or the DM is satisfied about results obtained.

Futhermore, the FITradeoff allows to visualize the performances of alternatives against each criterion, after the ranking and in the meanwhile the DM answers to the questions, discarding step by step those alternatives not anymore suitable according to the "weight assigned" and to the questions answered.

3.5 Conclusions

The results of this application are described by Oppio et al. (2016) and Dell'Ovo et al. (2017, 2018b). What emerges by this case study is that there are no good or bad aggregation procedures, but according to the context and to the data available, it is possible to recognise the most suitable one.

In detail, the AHP could incur in the problem of cognitive overload since the DMs are called to express their intensity of preference by pair comparison according to a semantic scale. Thus, internal consistency of judgements has to be checked and this is one of the most common critical issue (Whitaker 2007). For what concerns the Weighted sum, since it is a compensatory method, bad performances are compensated by good ones. This is not necessarily a critical issue, but it is important to evaluate the case study in order to understand if there are specific requirements or thresholds to respect. Despite the FITradeoff method is interactive and flexible, it shows some common difficulties within the second step. The number of questions a DM is called to answer is directly proportional to the number of criteria and their rank. Higher the number of criteria and higher could be the number of questions to be answered. Even if the set of criteria has been reviewed and the number decreased, the FITradeoff method doesn't allow to create a decision tree and breakdown the problem into a hierarchy.

This contribution has been a field of experimentation able to demonstrate pros and cons of the adopted methodologies. According to the literature review (see Chap. 1), some approaches have been selected and then applied to the decision problem concerning the location of healthcare facilities. The strength of this step is not given by the methodological choices, but by a deep understanding of the field of application and of the decision context that is crucial for developing a MCDA process.

Another important issue came to light after this testing phase. Here a finite set of alternatives have been analysed and evaluated the most suitable one among those present, but are we sure it is good enough? Or maybe is it only the best of a bad bunch? Are we sure about our decision? Are there other alternatives that we did not consider?

These methodologies allow to evaluate a finite set of alternatives but what if we do not have a limited number of option? What happen when the space of location alternatives is expanded to the whole territory?

Given the spatial nature of location problems, the integration with GIS is essential in order to evaluate the most suitable location for healthcare facilities and to expand the decision domain without being focused on a restricted number of alternatives.

References

Abdullahi S, Mahmud ARB, Pradhan B (2014) Spatial modelling of site suitability assessment for hospitals using geographical information system-based multicriteria approach at Qazvin city, Iran. Geocarto Int 29(2):164–184

Beheshtifar S, Alimoahmmadi A (2015) A multiobjective optimization approach for location-allocation of clinics. Int Trans Oper Res 22(2):313–328

Bottero MC, Buffoli M, Capolongo S, Cavagliato E, di Noia M, Gola M, Speranza S, Volpatti L (2015) A multidisciplinary sustainability evaluation system for operative and in-design hospitals. In: Improving sustainability during hospital design and operation. Springer, Cham, pp 31–114

Bottero MC, Caprioli C, Berta M (2020) Urban problems and patterns of change: the analysis of a downgraded industrial area in Turin. In: Values and functions for future cities. Springer, Cham, pp 385–401

Bottomley PA, Doyle JR, Green RH (2000) Testing the reliability of weight elicitation methods: direct rating versus point allocation. J Mark Res 37(4):508–513

Bourne L (2005) Project relationship management and the stakeholder circle. Doctor of Project Management. Graduate School of Business

BRE Global Ltd (2010) Breeam healthcare. BRE, Watford

Buffoli M, Nachiero D, Capolongo S (2012) Flexible healthcare structures: analysis and evaluation of possible strategies and technologies. Ann Ig 24(6):543–552

Burkey ML (2012) Decomposing geographic accessibility into component parts: methods and an application to hospitals. Ann Reg Sci 48(3):783–800

Burkey ML, Bhadury J, Eiselt HA (2012) A location-based comparison of health care services in four US states with efficiency and equity. Socio-Econ Plann Sci 46(2):157–163

Capolongo S, Sdino L, Dell'Ovo M, Moioli R, Della Torre S (2019) How to assess urban regeneration proposals by considering conflicting values. Sustainability 11(14):3877

Cergas SB (2017) Rapporto OASI 2017. Osservatorio sulle Aziende e sul Sistema sanitario Italiano

Chiu JE, Tsai HH (2013) Applying analytic hierarchy process to select optimal expansion of hospital location: the case of a regional teaching hospital in Yunlin. In: 2013 10th international conference on service systems and service management. IEEE, pp 603–606

Crivellini M, Galli M (2011) Sanità e salute: due storie diverse: sistemi sanitari e salute nei paesi industrializzati. F. Angeli

Daskin MS, Dean LK (2005) Location of health care facilities. In: Operations research and health care. Springer, Boston, pp 43–76

de Almeida AT, Almeida JA, Costa APCS, Almeida-Filho AT (2016) A new method for elicitation of criteria weights in additive models: flexible and interactive tradeoff. Eur J Oper Res 250(1):179–191

Dell'Ovo M, Capolongo S (2016) Architectures for health: between historical contexts and suburban areas. Tool to support location strategies. TECHNE-J Technol Architect Environ, 269–276

Dell'Ovo M, Capolongo S, Oppio A (2018a) Combining spatial analysis with MCDA for the siting of healthcare facilities. Land Use Policy 76:634–644

Dell'Ovo M, Frej EA, Oppio A, Capolongo S, Morais DC, de Almeida AT (2018b) FITradeoff method for the location of healthcare facilities based on multiple stakeholders' preferences. In: International conference on group decision and negotiation. Springer, Cham, pp 97–112

Dell'Ovo M, Frej EA, Oppio A, Capolongo S, Morais DC, de Almeida AT (2017) Multicriteria decision making for healthcare facilities location with visualization based on FITradeoff method. In: International conference on decision support system technology. Springer, Cham, pp 32–44. https://doi.org/10.1007/978-3-319-57487-5_3

Dente B (2014) Understanding policy decisions. In: Understanding policy decisions. Springer, Cham, pp 1–127

Du G, Sun C (2015) Location planning problem of service centers for sustainable home healthcare: evidence from the empirical analysis of Shanghai. Sustainability 7(12):15812–15832

Faruque LI, Ayyalasomayajula B, Pelletier R, Klarenbach S, Hemmelgarn BR, Tonelli M (2012) Spatial analysis to locate new clinics for diabetic kidney patients in the underserved communities in Alberta. Nephrol Dial Transplant 27(11):4102–4109

Fishburn PC (1967) Additive utilities with incomplete product set: applications to priorities and assignments. Operations Research Society of America (ORSA), Baltimore

Gamboa G, Munda G (2007) The problem of windfarm location: a social multi-criteria evaluation framework. Energy Policy 35(3):1564–1583

Gola M, Fugazzola E, Rebecchi A (2018) Mapping and programming healthcare services for new health perspectives. In: Healthcare facilities in emerging countries. Springer, Cham, pp 89–111

Gola M, Settimo G, Capolongo S (2019) Indoor air quality in inpatient environments: a systematic review on factors that influence chemical pollution in inpatient wards. J Healthc Eng 8358306

Janssen R, Van Herwijnen M, Beinat E (2000) Definite for windows. A system to support decisions on a finite set of alternatives (Software and package and user manual)

Keeney RL (1992) Value focused thinking. Harvard University Press, Cambridge

Keeney RL, Raiffa H (1976) Decisions with multiple objectives: preferences and value trade-offs. Cambridge University Press, Cambridge

Kim JI, Senaratna DM, Ruza J, Kam C, Ng S (2015) Feasibility study on an evidence-based decision-support system for hospital site selection for an aging population. Sustainability 7(3):2730–2744

Lee KS, Moon KJ (2014) Hospital distribution in a metropolitan city: assessment by a geographical information system grid modelling approach. Geospatial Health, 537–544

Mendelow AL (1981) Environmental scanning-the impact of the stakeholder concept. In: ICIS, p 20

Murad AA (2005) Using GIS for planning public general hospitals at Jeddah City. Environ Des Sci 3(3):22

Murad AA (2007) Creating a GIS application for health services at Jeddah city. Comput Biol Med 37(6):879–889

Noon CE, Hankins CT (2001) Spatial data visualization in healthcare: supporting a facility location decision via GIS-based market analysis. In: Proceedings of the 34th annual Hawaii international conference on system sciences. IEEE, p 10

Oppio A, Buffoli M, Dell'Ovo M, Capolongo S (2016) Addressing decisions about new hospitals' siting: a multidimensional evaluation approach. Ann dell'Istituto superiore di sanità 52(1):78–87

Rebecchi A, Boati L, Oppio A, Buffoli M, Capolongo S (2016) Measuring the expected increase in cycling in the city of Milan and evaluating the positive effects on the population's health status: a community-based urban planning experience. Ann Ig 28(6):381–391

Roy B (1985) Multicriteria methodology for decision analysis. Kluwer Academic Publishers

Roy B (2005) Paradigms and challenges. In Multiple criteria decision analysis: state of the art surveys. Springer, New York, pp 3–24

Saaty RW (1987) The analytic hierarchy process—what it is and how it is used. Mathe Modell 9(3–5):161–176

Saaty TL (1980) The analytic hierarchy process. Mc-Graw-Hill, New York

Shariff SR, Moin NH, Omar M (2012) Location allocation modeling for healthcare facility planning in Malaysia. Comput Ind Eng 62(4):1000–1010

Soltani A, Marandi EZ (2011) Hospital site selection using two-stage fuzzy multi-criteria decision making process. J Urban Environ Eng 5(1):32–43

Triantaphyllou E (2000) Multi-criteria decision making methods. In: Multi-criteria decision making methods: a comparative study. Springer, Boston, pp 5–21

USGBC (2011) LEED for healthcare. USGBC, Washington

Vahidnia MH, Alesheikh AA, Alimohammadi A (2009) Hospital site selection using fuzzy AHP and its derivatives. J Environ Manage 90(10):3048–3056

Whitaker R (2007) Criticisms of the analytic hierarchy process: why they often make no sense. Math Comput Model 46(7–8):948–961

Wu CR, Lin CT, Chen HC (2007) Optimal selection of location for Taiwanese hospitals to ensure a competitive advantage by using the analytic hierarchy process and sensitivity analysis. Build Environ 42(3):1431–1444

Wu WH, Lin CT, Peng KH, Huang CC (2012) Applying hierarchical grey relation clustering analysis to geographical information systems–a case study of the hospitals in Taipei City. Expert Syst Appl 39(8):7247–7254

Yang XS (2014) Nature-inspired optimization algorithms. Elsevier

Zhang P, Ren X, Zhang Q, He J, Chen Y (2015) Spatial analysis of rural medical facilities using huff model: a case study of Lankao county, Henan province. Int J Smart Home 9(1):161–168

Chapter 4
Modelling the Spatial Decision Problem. Bridging the Gap Between Theory and Practice: SitHealth Evaluation Tool

Abstract Decision Support Systems (DSSs) have been developed for aiding Decision-Makers (DMs) in handling and solving complex problems. When decisions belong to the public domain, as for the location of healthcare facilities, the use of DSSs could be fundamental to justify choices made and to easily communicate the results to citizens. Given the nature of the decision problem, the support of the Multicriteria-Spatial Decision Support Systems (MC-SDSS) has been recognised as strategic given the possibility to frame the decision problem, by breaking it down into hierarchical levels and to visualise the result directly on the territory through the use of maps. The contribution aims at presenting the strengths and weaknesses of the MC-SDSS for the location of healthcare facilities tested on a case study located in the Municipality of Milan, Italy, and to propose a DSS developed to support DMs, namely the Evaluation Tool for the Siting of Healthcare facilities (SitHealth Evaluation Tool).

4.1 Introduction

Facing problems which involve different dimensions and several stakeholders require multi-disciplinary knowledges and usually the human mind is poorly equipped to solve it unaided (Keeney 1992). In this context, Decision Support Systems (DSSs) can support Decision-Makers (DMs) in handling and solving the problem. DSSs are a group of softwares that provide help to one or more steps of decision-making and are able to enhance its processes (Anwar and Ashraf 2014). In detail DSSs are able to improve the quality of decisions by justifying choices made with the support of tools and technologies (Assumma et al. 2020; Brambilla et al. 2019; Brambilla and Capolongo 2019). Figure 4.1 shows the typical structure of DSSs where data are collected and then processed by softwares in order to result in appropriate outputs.

M. Dell'Ovo et al., *Decision Support System for the Location of Healthcare Facilities*,
PoliMI SpringerBriefs, https://doi.org/10.1007/978-3-030-50173-0_4

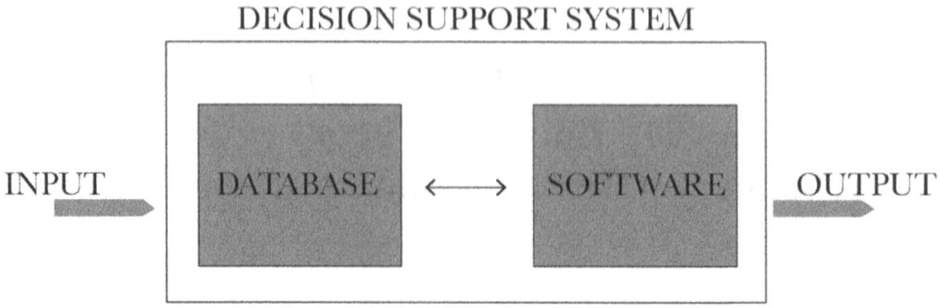

DECISION SUPPORT SYSTEM

INPUT   DATABASE ⟷ SOFTWARE   OUTPUT

Fig. 4.1 Structure of DSS. Adapted from Anwar and Ashraf (2014)

They can be interactive and flexible, learning by doing is the keyword of these softwares, since they can be continuously implemented and improved considering emerging needs from the context under investigation. They are used in different fields (health, economy, architecture, planning, agriculture etc.). Given the user-friendly layout and interface, DSSs can be easily managed from non-expert DMs, who according to the specific phases of the process, become aware about the system and the decision they are taking.

When decisions belong to the public domain, the use of DSSs could be fundamental and strategic since they can justify choices made and easily communicate the results to citizens (Bottero et al. 2020). Moreover, they can aid DMs to a better understanding of the problem, to address decisions step by step, by the support of robust methodological frameworks. The process therefore results transparent in all its parts.

Within the context of healthcare facilities' location, it has been recognised how the Multicriteria-Spatial Decision Support Systems (MC-SDSS) (Malczewski and Rinner 2015) is strategic to frame the decision problem, by breaking it down into hierarchical levels to be better analysed by territorial maps. Since the location problem involves spatial information, Geographic Information System (GIS) can aid DMs to recognize new suitable alternatives, thus expanding the decision domain (Faroldi et al. 2019; Dell'Ovo et al. 2018a; Oppio et al. 2016).

Given these premises, the aim of the contribution is to describe the potentials and limits of MC-SDSS for the location of healthcare facilities and to introduce a DSS developed to support DMs, namely the Evaluation Tool for the Siting of Healthcare facilities (SitHealth Evaluation Tool).

The paper is divided into five sections. After a brief introduction aimed at describing the main characteristics of DSSs, Sect. 4.2 will present an application of MC-SDSS on a case study and main findings and criticalities will be discussed in Sect. 4.3. Section 4.4 is devoted at defining the evaluation tool developed and main conclusions are presented in Sect. 4.5.

4.2   Modelling the Spatial Problem

Given the spatial nature of the problem and the criteria involved, (see Chaps. 2 and 3), the problem concerning the location of healthcare facilities has been solved by applying the MC-SDSS. The multi-methodological evaluation framework defined in this research aims to address DMs in the process of hospitals new site selection and it has been applied in the Municipality of Milan, Italy to the case study of "La Città della Salute" (see Chap. 3).

4.2.1   How to Be Supported by MC-SDSS

Figure 4.2 shows the multi-methodological flowchart developed to solve the decision problem which allows DMs to understand from the beginning, how it is possible to face and manage complex choices. The framework mainly consists of four parts:

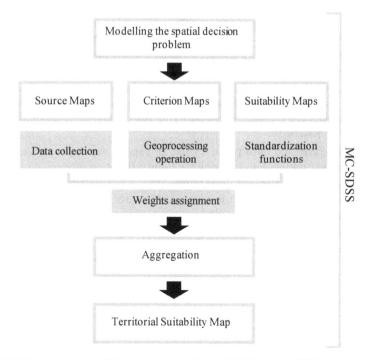

Fig. 4.2 Multi-methodological flowchart. Adapted from Dell'Ovo et al. (2018a)

- Definition of a set of criteria able to describe the problem in all its part (Roy 1985, 2005);

- Collection of spatial information corresponding to the set of criteria previously identified in order to perform all the geoprocessing operations for creating Source Maps, Criterion Maps and, after the standardization, the Suitability Maps.
- Criteria weights elicitation to assign a different importance to the criteria (Riabacke et al. 2009);
- Aggregation of obtained results and weights assigned to create the Overall Suitability Map concerning the location of healthcare facilities.

The first phase involves the definition of a consistent set of criteria, which have been structured considering a deep analysis of the literature review, existing evaluation tools, stakeholders' opinions and reviewed according to the suggestion received after the interaction with experts (see Chaps. 1, 2 and 3). For what concerns the second phase, it is important to underline how data sources are diverse; they could be collected from different databases. Within this context they mainly have been detected from the website of the Municipality of Milan, Geoportale of Lombardy Region, Open Street Map, Google maps and ISTAT. The operations developed in the GIS domain have been performed by the support of ArcGIS and in detail by using the "Spatial Analyst toolbox". Once data have been collected (Source Map) they can be processed. ArcGIS allows to perform different spatial operations in relation to the type of information required—Euclidean Distance; Slope; Density; etc.

Through these geoprocessing operations Source Maps are transformed into raster maps (Criterion Maps), which already carry specific information as interval, classes or ranges. Standardization is then a fundamental phase to set different data with respect to a common measurement scale. It is essential to operate a reclassification on all those maps formerly built, based on 10 intervals, where 0 represent the lowest score and 10 the best one.

For what concerns the third phase a panel of experts has been interviewed and asked to elicit their preferences by using the Point Allocation method, while for the fourth one the Weighted Linear Combination, WLC has been performed (Dell'Ovo et al. 2018a). The result obtained is a unique map able to answer to the decision problem framed, giving also the possibility to understand the performance according to the whole set of criteria. The map gives an overall evaluation of the whole territory under analysis.

4.2.2 *Applying the MC-SDSS*

Once provided a general overview of operations required and the methodologies proposed, it is possible to investigate in detail the location of healthcare facilities and in particular of the case study "La Città della Salute" in Milan, Italy (Dell'Ovo et al. 2017, 2018a, b; Oppio et al. 2016).

Results of the investigation and an in-depth description of the whole process, including moreover an explanation about how data have been collected, processed, standardized and aggregated can be appreciated in Dell'Ovo et al. (2018a).

Within this context a synthesis of the main phases will presented, pointing out the attention on the validation phase and the sensitivity analysis.

From Figs. 4.3, 4.4, 4.5, 4.6, 4.7, 4.8, 4.9, 4.10, 4.11, 4.12, 4.13 and 4.14 a fact-sheet for each criterion is presented with the basic information about its relevance within the decision context, a description, the data source, the scale, the analysis processed, the standardization function and the visualization of the Source Map, Criterion Map and Suitability Map. The economic dimension has not been subjected to spatial analysis since information regarding the ownership, the cost and the current condition of the area, is confidential and not in the public domain. This investigation requires a more detailed evaluation and it could be developed in a second phase, once identified a limited and finite number of satisfactory sites. In fact, data used within this context are open source and they can easily be found. This is an additional key element of the framework, since each criterion belonging to the decision tree is public and every Municipality could run this analysis on its own territory.

AIM

Preferring peripheral and degraded areas (DM 12/12/00).
Degraded areas are those one affected by the action of
natural factors or areas degraded by the action of anthro-
pogenic factors, as contaminated areas or in the presence
of abandoned factories.

DESCRIPTION

The criterion represents abandoned areas with information
regarding the land use, year of disposal, possible use after
disposal, degree of preservation.

DATA SOURCE

Abandoned areas map of the Lombardy Region (Geoportale Regione Lombardia)

SCALE

1:10000

ANALYSIS

Euclidean Distance

STANDARDIZATION

Distances ≤500 m are standardized to 10. Distances between 500 and the max Distances are standardized according to the linear score (the lower the Distances, the
higher the score). Ranges have been defined every 1000
m.

Source Map

Criterion Map

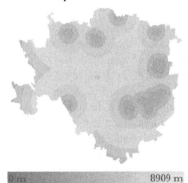

0 m   8909 m

Suitability Map

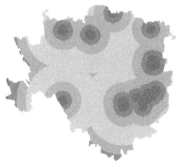

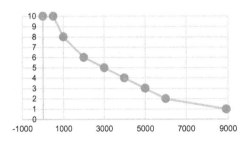

0 1 2 3 4 5 6 7 8 9 10

Fig. 4.3 Centre of urban redevelopment

AIM
By choosing an area with an appropriate land use, it is possible to handle more correctly its management and development. Moreover, it is possible to consider the availability of spaces for future expansions depending on the needs and requirement of the society.

DESCRIPTION
The criterion represents the different types of land cover.

DATA SOURCE
Land use map of the Municipality of Milan (Geoportale comune Milano)

SCALE
1:50000

ANALYSIS
Reclassify

STANDARDIZATION
Scores have been defined according to the land use, areas more suitable as the abandoned one are standardized to 10 while the unsuitable one as protected sites are standardized to 0.

No value function is provided since the values (0-10) have been assigned manually considering the land use present.

Source Map

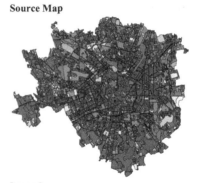

Legend

Suitability Map

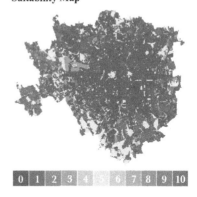

Fig. 4.4 Flexibility

AIM

Ability to promote sustainable development and to locate health services considering the population and its age. In fact, the criterion takes into consideration the catchment area of the hospital.

DESCRIPTION

The criterion aggregates information related to the population distribution and percentage of people older than 65 years.

DATA SOURCE

Istat: resident population and age of the population (2014)

SCALE

ANALYSIS

Reclassify

STANDARDIZATION

Scores have been defined according to the population resident in each district of the municipality of Milan and the percentage of people > 65 years.

No value function is provided since the values (0-10) have been assigned manually considering two factors: population and year of the population

Source Map

Municipio	Population	% > 65 y.o.
1	90.944	26
2	128.926	24
3	131.709	27
4	145.260	28
5	113.638	27
6	137.308	30
7	158.025	29
8	168.456	29
9	167.602	26

Suitability Map

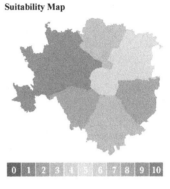

0 1 2 3 4 5 6 7 8 9 10

Fig. 4.5 Building density

AIM

Promoting the accessibility of the hospitals by considering both private and public mobility and moreover the presence of parking lots.

DESCRIPTION

The criterion aggregates information related to the public mobility concerning bus stops and metro stop.

DATA SOURCE

Subway stops and bus stops map (Geoportale comune Milano

SCALE

1:5000

ANALYSIS

Euclidean Distance

STANDARDIZATION

Public Mobility

Bus: Distances ≤400 m are standardized to 10. Distances between 400 m and 1500 m are standardized according to the linear score (the lower the Distances, the higher the score). Distances > 1500 m are standardized to 3 (15 min walking Distances).

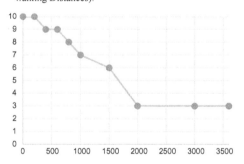

Metro: Distances ≤ 800 m are standardized to 10. Distances between 800 and 2000 are standardized according to the linear score (the lower the Distances, the higher the score). Distances between 2000 m and 3000 m are standardized to 3 (30 min walking Distances). Distances between 3000 m and the max Distances are standardized to 1.

Source Map
Public Mobility:
- Bus

- Metro

Criterion Map
Public Mobility

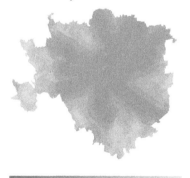

0 m   8304 m

Fig. 4.6 Accessibility (public mobility)

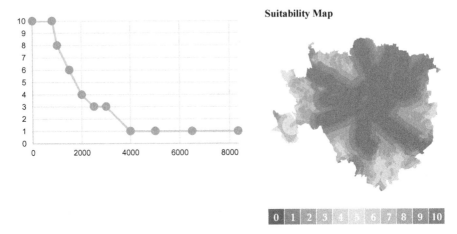

Fig. 4.6 (continued)

AIM
Promoting the accessibility of the hospitals by considering both private and public mobility and moreover the presence of parking lots.

DESCRIPTION
The criterion aggregates information related to the highway and fast road.

DATA SOURCE
Highway and primary road map (Geoportale comune Milano)

SCALE
1:5000

ANALYSIS
Euclidean Distance

STANDARDIZATION
Private Mobility
Distances ≤500 m are standardized to 10. Distances between 500 m and 3000 m (15 min car Distances) are standardized according to the linear score (the lower the Distances, the higher the score) until the score 6. Distances between 3000 m and the 5500 m are standardized from 3 to 1. Distances ≥5500 m are standardized to 0.

Source Map

Criterion Map

0 m   5870 m

Suitability Map

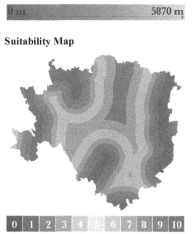

0 | 1 | 2 | 3 | 4 | 5 | 6 | 7 | 8 | 9 | 10

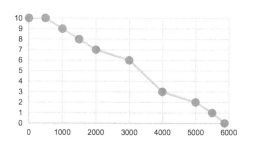

Fig. 4.7 Accessibility (private mobility)

AIM

Promoting the accessibility of the hospitals by considering both private and public mobility and moreover the presence of parking lots.

DESCRIPTION

The criterion aggregates information related to the availability of parking lots.

DATA SOURCE

Parking lot map (Geoportale comune Milano)

SCALE

1:5000

ANALYSIS

Euclidean Distance

STANDARDIZATION

Parking

Distances ≤800 m are standardized to 10. Distances between 800 m and 2000 m are standardized according to the linear score (the lower the Distances, the higher the score). Distances between 2000 and 3000 (30 min walking Distances) are standardized to 3. Distances between 3000 and 7000 are standardized to 1. Distances ≥7000 m are standardized to 0.

Source Map

Criterion Map

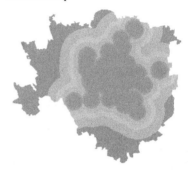

0 m   9620 m

Suitability Map

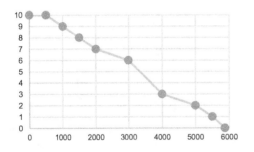

| 0 | 1 | 2 | 3 | 4 | 5 | 6 | 7 | 8 | 9 | 10 |

Fig. 4.8 Accessibility (parking lot)

AIM
Promoting the presence of services in order to avoid the isolation of the hospital and to promote the integration within the surrounding context

DESCRIPTION
The criterion aggregates main important services in the Municipality of Milan

DATA SOURCE
Shopping mall, shops, school, churches, healthcare facilities map (Geoportale Regione Lombardia)

SCALE
1:10000

ANALYSIS
Euclidean Distance

STANDARDIZATION
Distances ≤800 m are standardized to 10. Distances between 800 m and 7000 m are standardized according to the linear score (the lower the Distances, the higher the score). Distances ≥7000 m are standardized to 0.

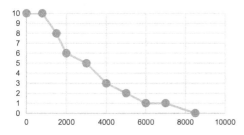

Source Map

Criterion Map

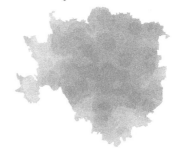

8500 m

Suitability Map

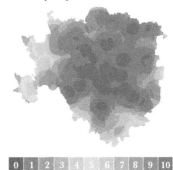

0 1 2 3 4 5 6 7 8 9 10

Fig. 4.9 Services

AIM

Supporting the selection of an area surrounded by a natural environment. Preserving the natural environment to future changes and developments.

DESCRIPTION

The criterion represents the presence of parks, garden and green spaces.

DATA SOURCE

Green areas map (geoportale Regione Lombardia)

SCALE

1:1000

ANALYSIS

Euclidean Distance

STANDARDIZATION

Distances ≤ 500 m are standardized to 10. Distances between 500 m and 1500 m (15 min walking Distances) are standardized according to the linear score (the lower the Distances, the higher the score). Distances between 1500 m and 3000 m are standardized to 3. Distances between 3000 m and the max Distances are standardized to 1.

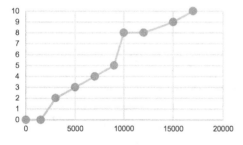

Source Map

Criterion Map

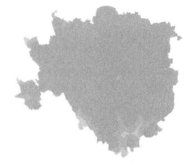

0 m   5700 m

Suitability Map

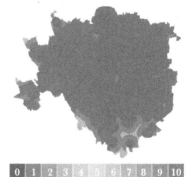

0 1 2 3 4 5 6 7 8 9 10

Fig. 4.10 Green area

AIM
Selecting an area close to the collecting system and to a vacuum sewer station or presence to an efficient network system.

DESCRIPTION
The criterion has been represented considering the sewerage system

DATA SOURCE
PUGGS of the Municipality of Milan

SCALE

ANALYSIS
Euclidean Distance

STANDARDIZATION
Distances ≤ 800 m are standardized to 10. Distances between 800 m and 5000 m are standardized according to the linear score (the lower the Distances, the higher the score). Distances between 5000 m and 7000 m are standardized to 1. Distances ≥ 7000 m are standardized to 0.

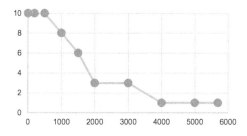

Source Map

Criterion Map

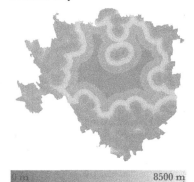

0 m   8500 m

Suitability Map

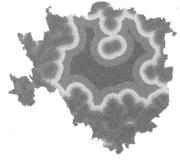

Fig. 4.11 Network infrastructures

AIM
Observing the regulation concerning the noise pollution in order to respect the low and avoid future potential problems.

DESCRIPTION
The criterion represents the acoustic classifications approval status.

DATA SOURCE
Acoustic plans map (Geoportale comune Milano)

SCALE
1:10000

ANALYSIS
Reclassify

STANDARDIZATION
Scores have been defined according to the acoustic classification defined for the Municipality of Milan and the max emission regulated by the law.
0 dB is standardized to 10;
50 dB are standardized to 10;
55 dB are standardized to 7;
60 dB are standardized to 5;
65 dB are standardized to 3;
70 dB are standardized to 1.

No value function is provided since the values (0-10) have been assigned manually considering acoustic classification.

Source Map

Legend
- 0
- 50
- 55
- 60
- 65
- 70

Suitability Map

0 1 2 3 4 5 6 7 8 9 10

Fig. 4.12 Noise pollution

AIM

Preventing health issue by considering the level of air pollution within the context where the area is located.

DESCRIPTION

The criterion aggregates information related to the level of pollution of three different pollutants: PM10, O3, NO2.

DATA SOURCE

ARPA Lombardia: PM10, O3, NO2 annual average (January 2016-December 2016)

SCALE

ANALYSIS

Reclassify

STANDARDIZATION

Scores have been defined according to the level of pollutants present in each district of the Municipality of Milan according to the position of detected stations.

No value function is provided since the values (0-10) have been assigned manually considering presence of air pollutants.

Source Map

position of detection stations in the Municipality of Milan

Suitability Map

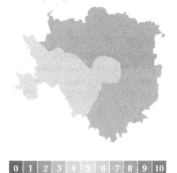

0 1 2 3 4 5 6 7 8 9 10

Fig. 4.13 Air pollution

AIM

Preventing health issue by identified unhealthy issue and maintaining a safe distance. For unhealthy industry are defined the manufactures or factories that produce vapours, gases or other unhealthy fumes that are dangerous for the health of inhabitants.

Source Map

DESCRIPTION

The criterion represents the presence of industrial activities

DATA SOURCE

Location of Technological Network buildings map (Geoportale comune Milano)

SCALE

1:1000

Criterion Map

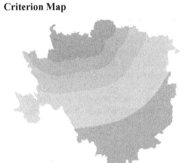

ANALYSIS

Euclidean Distance

STANDARDIZATION

Distances ≤1500 m are standardized to 0. Distances between 1500 m and 15000 m are standardized according to the linear score (the higher the Distances, the lower the score). Distances ≥15000 m are standardized to 10.

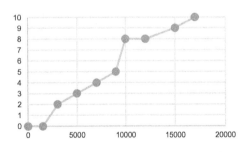

0 m   8500 m

Suitability Map

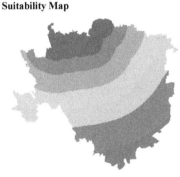

| 0 | 1 | 2 | 3 | 4 | 5 | 6 | 7 | 8 | 9 | 10 |

Fig. 4.14 Unhealthy industries

Moreover, some important prerequisites have been added to the framework and they can be evaluated as constraints which can strongly influence the final decision since areas affected by them will not be any more judged as suitable. The additional layers selected are: existing hospitals and areas of hydrological and hydraulic instability.

For what concerns the standardization, the definition of intervals/ranges for each criterion and the subsequent allocation of values (0–10) have been addressed by comparing experts' opinions, analysing the Italian regulations and laws as well as the literature. In detail the standardization function of each criterion can be appreciated in the fact-sheets.

Moreover, the tool allows to visualize a table for each Suitability map able to elicit the number of cells present in the map scoring with different performances (from 0 to 10). Figure 4.15 shows the median value scored by each criterion.

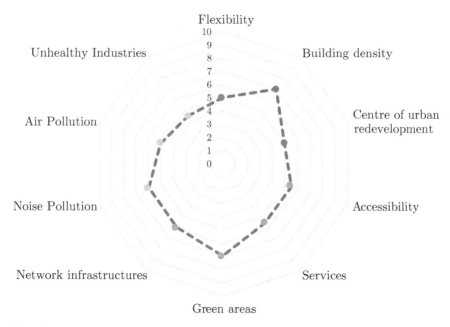

Fig. 4.15 Median value scored according to the set of criteria

Once the criteria have been standardized, they can be aggregated. Considering the weights previously elicited (See Chap. 3, Table 3.8), the aggregation has been carried out according to the WLC. Since the decision tree is composed by multiple dimensions, it is more interesting to analyse the partial suitability maps before the Overall Suitability map (Fig. 4.16), in order to reflect on the partial results obtained by each area, the median value performed (Fig. 4.17) and also the value performed by the pre-selected sites for the location of "La Città della Salute".

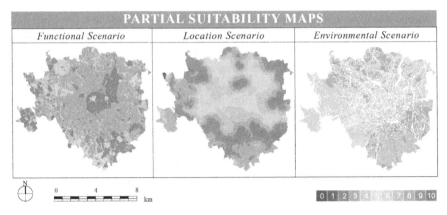

Fig. 4.16 Partial Suitability maps. Adapted from Dell'Ovo et al. (2018)

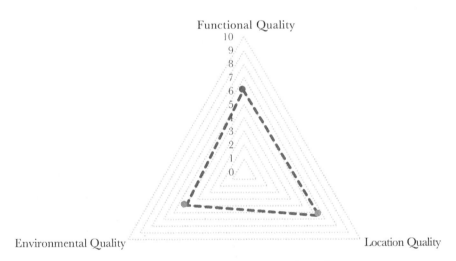

Fig. 4.17 Median value scored by the aggregation within each dimension

Since this analysis represents a test of the methodology, only the areas located in the Municipality of Milan have been studied, thus disregarding the site present in "Sesto San Giovanni" which is located in another urban context.

Trying to interpret the results obtained:

- Functional Quality: sites located in the centre perform a lower score compared to those ones located in the suburban areas;
- Location Quality: also in this case the suburban areas perform a higher score since more suitable under the private mobility point of view;
- Environmental Quality: it is evident how the map is divided in three main zones and the most suitable is located on the south-east side of the Municipality.

The Overall Suitability Map (Fig. 4.18) is obtained by the aggregation of the partial results previously described and presented. What emerges from this first attempt is that the maximum value scored by the sites located in the Municipality of Milan is 8, no sites got 10 and the median value is 5.5. It means that even if it is possible to identify some suitable areas, no one can be considered as the optimal one according to the decision tree structured and the set of weights assigned. Moreover, Fig. 4.18 overlaps the results obtained by the prerequisites selected (i.e. existing hospitals and areas under hydrological and hydraulic instability). The first prerequisite has been chosen in order to equally distribute another hospital on the territory, while the second to highlight river banks and areas subjected to phenomena such as landslides, floods. The last map underlines the location of five of the six areas identified by the Lombardy Region across the real decision process for the location of "La Città della Salute".

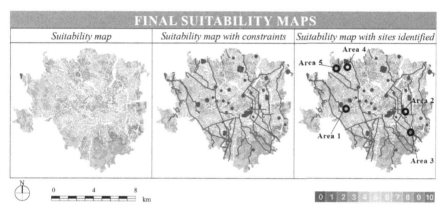

Fig. 4.18 Final suitability maps with and without prerequisites and with potential candidate sites for the location of "La Città della Salute". Adapted from Dell'Ovo et al. (2018)

The last map presented in Fig. 4.18 shows how Site 2, Site 3 and Site 5 are located in areas of hydrological and hydraulic instability or nearby them, while Site 1 and Site 4 scored between 3 and 4.

Considering these first results, it is possible to underline how the visualization helps moreover to communicate information in a clear way also to non-experts. Another important key feature provided by the MC-SDSS is the possibility to generate new and unexpected alternatives (Colorni and Tsoukiàs 2018). By excluding a priori some sites and restricting the evaluation to a finite number, the possibility to make unsatisfactory choices arises.

4.3 Sensitivity Analysis and Main Findings

In order to simulate different scenarios, a Sensitivity analysis (Ferretti and Pomarico 2013) has been performed, by changing the weights previously defined by experts. In particular a "What if" analysis has been carried out, that allows to change the importance assigned to each dimension involved in the analysis. In this case it has been assigned the same influence to the level of the criteria and the macro-areas' weights (Functional Quality; Location Quality; Environmental Quality) have been changed, as it is illustrated in the Table 4.1.

Results obtained confirm what emerged from the Suitability maps presented in Fig. 4.18. As it has been already mentioned, there are no sites performing 10 as suitability score in the overall map (Fig. 4.19):

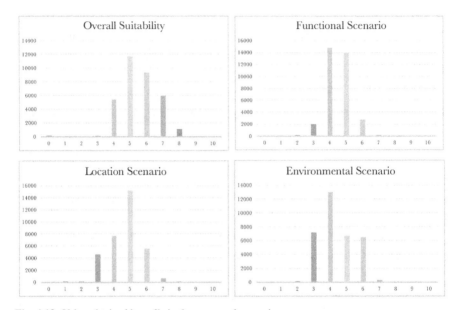

Fig. 4.19 Value obtained by cells in the proposed scenarios

- Functional Scenario: the maximum value obtained is 7 and the median value is 5;
- Location Scenario: the maximum value obtained is 8 and the median value is 5;
- Environmental Scenario: the maximum value obtained is 7 and the median value is 5.

The overall suitability median value in the perspectives ("What if" Analysis) visualized in Fig. 4.20 decreased of 0.5 and the five areas analysed confirmed their

Table 4.1 Weights assigned for the sensitivity analysis

		Functional scenario (%)	Location scenario (%)	Environmental scenario (%)
WEIGHTS	Functional quality	70	15	15
	Location quality	15	70	15
	Environmental quality	15	15	70

scores. In the Functional and Location Scenarios, sites located in the suburban areas have performed a higher score compared to the central one, since, in the first case, abandoned areas are mostly located in the outskirt of the city, while in the second case the private mobility affects mostly the border of the city (Rebecchi et al. 2019). For what concerns the Environmental Scenario, the result is different and sites located in the north-western part of the city performed a lower score, because of unhealthy industries present in that part of the city of Milan.

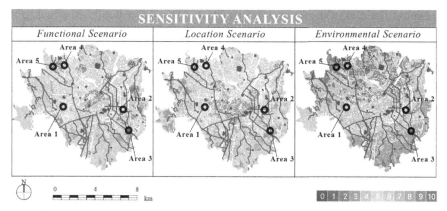

Fig. 4.20 Scenarios developed by the sensitivity analysis

The Sensitivity analysis allows to validate the results obtained and to control the internal robustness, thus adding objectivity to the process. According to the scenarios visualised in Fig. 4.20 and the results obtained in Fig. 4.18, it could be recommended to locate hospitals in the south-eastern part of the city of Milan, since is the most stable one and it has obtained a high score in the overall maps presented.

A critical issue detected from this first operational attempt is the exclusion from the spatial analysis of the Economic Dimension, postponed to a second and more detailed phase. The idea is that a further MCDA could be performed in order to evaluate a finite number of alternatives resulted as the most suitable from the MC-SDSS, thus including economic data.

It is also important to better define the role of the DMs in the process since a group of experts have been interviewed on the weights assignment. This phase strongly affects the final result. When applied to a real case study the sensitivity analysis could help to overcome this issue by checking the internal consistency and the robustness of the results.

This kind of evaluation belongs to the field of the "constructive evaluation since putting in evidence weakness and strengths of each area (pixel) it becomes a support for the decision making process by increasing moreover, its transparency" (Oppio et al. 2016).

The combination of MCDA and GIS has been fundamental in this decision context to solve this complex problem since it gave the possibility to consider simultaneously multiple dimensions and to show how each phase has performed step by step.

Moreover, the Sensitivity analysis allowed to verify the stability of the results by performing the "What if" Scenario, checking its internal robustness, thus legitimating the final decision and reducing subjectivity.

Given the complexity of the problem and the presence of both intrinsic and extrinsic characteristics, it is possible to outline how each context has its own territorial and social features. For this reason, it is not possible to provide a unique solution aimed at solving the location of healthcare facilities, but, by analysing the territorial and social context, a set of options among which DMs are free to decide can be provided.

Even if the role of the analyst is fundamental to support the development and comprehension of each phase, within this decision process, DMs are more conscious about their role and able to take a decision.

4.4 Evaluation Tool for the Siting of Healthcare Facilities (SitHealth Evaluation Tool)

The Evaluation Tool for the Siting of Healthcare facilities (SitHealth Evaluation Tool) is structured according to the four stages of the MCDA (see Chap. 2), namely Stakeholders analysis, Criteria definition, Criteria weight elicitation and Aggregation procedure.

This is a trial that needs to be further implemented and developed by high-level programming language (e.g. Python, Java, etc.), but it can be considered as a first attempt able to show how to combine the lesson learned from this first application and how to bridge the gap between theory and practice. The tool should work with open source GIS data provided by each Country, Region or Municipality and be connected to a GIS software (e.g. ArcGIS, QGIS, ILWIS, etc.) able to collect, manage and process this kind of spatial information.

For what concerns the four stages (stakeholder analysis, criteria definition, weights assignment, aggregation procedure), a solution is not given to the DMs but an interactive instrument that can be shaped according to their practical instances.

In order to use it properly, basic knowledges about decision making processes and MCDA are necessary. Even if it can be easily used also by non-experts, to have an appropriate result, the support of an analyst is suggested.

Below the main steps and characteristics of the tools area presented.

4.4.1 Operational Recommendations

The SitHealth Evaluation Tool is a DSS developed to aid public authorities, local administrations, private companies, hospitals' general managers in charge to find a suitable location for new hospitals or to relocate existing ones. It combines the methodology provided by MCDA with the potentials of GIS, allowing a spatial analysis. The SitHealth provides a deep knowledge about strengths and weaknesses of the territory investigated according to a set of criteria aimed to address DMs towards robust decisions about the location of healthcare facilities.

The startup screen of the tool provides a general description about its features, the potential users and purposes, the methodological framework and the main phases (Fig. 4.21).

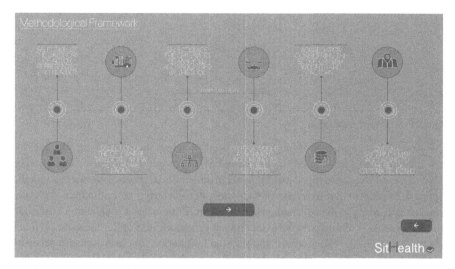

Fig. 4.21 Methodological framework

In detail, the first phase concerns the selection of the stakeholders, the second one the area where the hospital has to be located, the third one the criteria according to the context and actors' needs, the forth the weights' assignment, the fifth the choice of the procedure for aggregating scores and weights and, finally, in the sixth one the Suitability maps.

For what concerns the stakeholder represented, five categories of possible stakeholders are presented with a brief description. The categories considered are those defined by Dente (2014), namely Political Actors; Bureaucratic Actors; Actors with Special Interest; Actors with General Interest; Experts. In case more than one actor (Fig. 4.22) and category is involved, it is suggested to perform the processes a number of times equal to the number of actors involved and then to compare the results.

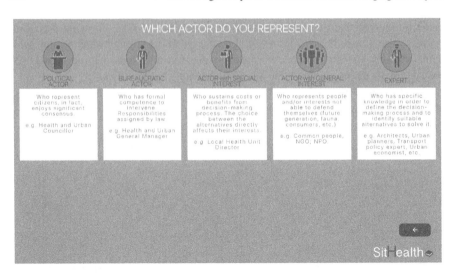

Fig. 4.22 Stakeholder selection

Before the criteria definition, it is required to select the city where the health-care facility has to be located, or the whole Province and the Region. The criteria listed in the software are those available according to the territorial context analysed. Figure 4.23 presents an example of fact-sheet completed to be completed for each criterion according to the following issues: why it is important to select this criterion, which elements it includes; where data have been taken; the scale; the type of analysis performed in the GIS domain and the standardization function (since in order to be aggregated, criteria have to be classified using the same scale). By selecting the button inside the text box of the standardization, it is possible to visualize the value function created which is fixed and cannot be modified. The DMs once understood the meaning of the criterion can decide to select or disregard it and use other criteria, in fact, the set of criteria can be tailored to the specific case study under investigation. When all the fact-sheets have been completed and the criteria adequate to the context selected, it is possible start the evaluation process.

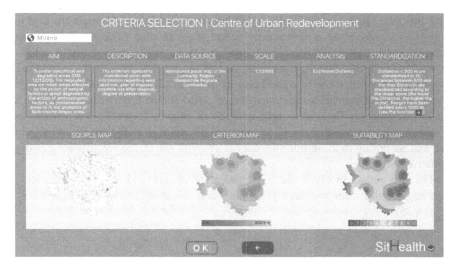

Fig. 4.23 Criteria fact-sheet

In the next phase the DMs can assign a different importance to the set of the selected criteria (Fig. 4.24).

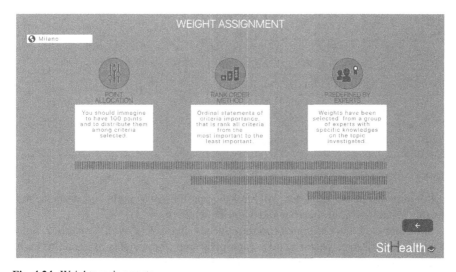

Fig. 4.24 Weights assignments

Three different procedures are proposed (Riabacke et al. 2012) and their use is related and suggested according to the category of actor represented.

The point allocation (Bottomley et al. 2000) allows to distribute 100 points among the set of criteria. It yields a cardinal rather that an ordinal scale of importance. This allocation is based strictly upon a DMs' judgements. The Rank Order Method allows to rank criteria with ordinal statement from the most important to the least important.

This procedure is suggested when the preference order is clear, or when the DM is not able to express the preference by specifying a precise value but only if one criterion is more preferred than another one and so on. When the number of criteria is high, the rank could be very complex, so it is suggested to proceed by comparing two by two the criteria until a final rank is obtained. The last method considers a Predefined set of weights assigned by experts with specific knowledge on the topics under investigation. By selecting the icon "Predefined by Experts", it is possible to visualize their judgments. Experts have assigned weights to all the criteria of the data set, and according to those ones the user is going to select, the weights are equally redistributed.

The last phase involves the aggregation procedure and within this context two aggregation rules are here presented: compensatory and non-compensatory (Fig. 4.25).

Fig. 4.25 Aggregation procedures

Running this phase, the DMs will have the possibility to obtain and visualise the Overall Suitability Map (Ferretti and Montibeller 2016).

Choosing the compensatory procedure means that a WLC will be applied, able to combine the scores performed by each pixel with the weights assigned to each criterion. Data are aggregated and a negative performance is compensated by a good performance, without any threshold. In case of non-compensatory aggregation techniques, thresholds of acceptability for each criterion should be introduced. This threshold levels (minimum performance) that "alternatives" have to meet, can be defined as a kind of preliminary requirement. If the level is not respected, alternatives can be excluded from the evaluation. This procedure allows to reject not suitable areas before the final aggregation. By performing this analysis, it is possible to strengthen the importance of some characteristics that the location of an hospital should have.

Once the most appropriate aggregation procedure has been selected, the software geo-processes all the information provided and it is possible to visualize the Overall Suitability Map (Fig. 4.26).

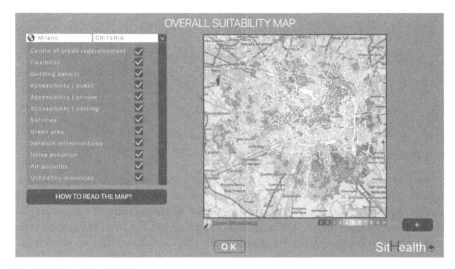

Fig. 4.26   Overall suitability map

Since this step could be not usual for non-expert users, an explanation of the meaning of the map is provided (see "How to read the map").

Before to conclude the session, in case the DMs want to test another aggregation procedure, or to change the weights assigned in order to check the robustness of the results obtained within the first evaluation round, they have the possibility to review the previous phases. By changing the weights, it is possible to visualise different scenarios and to validate the results obtained by checking their stability with respect to changing inputs. Some operational recommendations are provided (Fig. 4.27) to better understand the partial outputs of each phase and their relevance within the decision process.

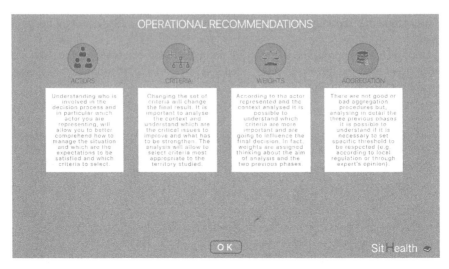

Fig. 4.27 Operational recommendations

Once the session is concluded, in case of multi-stakeholders, it is suggested to perform again the whole analysis in order to compare the different Suitability Maps obtained and to interpret them to achieve and take a shared decision. Otherwise the session is concluded and the DMs can export and review the obtained results. The graphs allow to communicate the results in an easy and immediate way, even for non-experts.

4.5 Conclusions

The research proposed aims to solve the complex decision problem of healthcare facilities location.

With the purpose to answer to the questions regarding the meaning of suitability, the research has deeply developed the four stages of the MCDA: the first about who is involved in the decision process—Stakeholder Analysis—(Who is in charge to define if an area is suitable or not?); the second one about the criteria to consider—Criteria definition—(Which characteristics are required to be suitable?), the third and the fourth about respectively the elicitation of criteria relevance—Criteria Weight Elicitation—and how to obtain a consistent result—Aggregation Procedure—(What is suitable?).

Trying to synthetize the main phases of the research, the necessity to be supported by a DSS for the location of healthcare facilities results quite clear. In fact, there are too many variables that cannot be generalised and controlled but need to be analysed. A predefined tool is not able to catch all the peculiar features of different territorial contexts, as well to answer to the stakeholders' specific needs.

Starting from the first step, the most important consideration resulted by the stakeholders' investigation, namely the importance to open the process to all the categories of actors who are affected by or able to influence the decision. Disregarding some of them and the role they play means not considering all the points of view, their expectations and therefore solving only partially the problem.

About the criteria definition, it is strictly connected to the actors involved and the territorial context. Since criteria are based on stakeholders' needs and expectations and the analysis of territorial strengths and weakness, they cannot be defined without these preliminary studies.

The weights assignment is also strongly connected to how the problem has been structured and who plays the role of the DM.

Finally, the choice of the aggregation procedure depends on all the previous phases, being the impact of compensatory or non-compensatory procedures relevant on the final result. Given these premises, the SitHealth Evaluation Tool has been developed with the aim of both supporting the DM in this complex decision and generating options instead of solutions. It has been defined as a support to guide the DMs across the overall decision process.

The SitHealth Evaluation Tool guides the DMs through the four stages until the definition of a suitability map. Different methodologies are provided for each phase and also an explanation for deciding which one is the most appropriate. The flexibility can be defined as the strength of the tool since it can be adapted according to specific instances emerging from time to time. Moreover, another important characteristic that deserves to be mentioned is its iterative nature, being possible to select those steps which need to be reviewed in case of mistakes or unsatisfactory results.

Each step is explained and graphically presented in order to facilitate its comprehension. The results obtained are visualized in an Overall Suitability Map able to aggregate the choices made and the elicited preferences. Moreover, the Partial Suitability Maps, in the fact-sheet of each criterion, allow to better understand under which aspect different areas result lacking or critical.

As it has been already briefly explained, the SitHealth Evaluation Tool is not only able to evaluate the location suitability but also to support design strategies. The tool can be used both to generate new suitable alternatives, if they have not been yet defined, and to understand the level of suitability of the given alternatives. In this second case, the Partial Suitability Maps allow to understand the reasons of critical results in case of ranking low alternatives. If these critical issues can be modified, the tool aids also to identify which design actions are more urgent and effective. Given the flexibility of the tool and its iterative nature, it is possible to test whether changes to the critical issues lead to better results.

References

Anwar N, Ashraf I (2014) Significance of decision support systems

Assumma V, Bottero M, Datola G, De Angelis E, Monaco R (2020) Dynamic models for exploring the resilience in territorial scenarios. Sustainability 12(1):3

Bottero M, Caprioli C, Berta M (2020) Urban problems and patterns of change: the analysis of a downgraded industrial area in Turin. In: Values and functions for future cities. Springer, Cham, pp 385–401

Bottomley PA, Doyle JR, Green RH (2000) Testing the reliability of weight elicitation methods: direct rating versus point allocation. J Mark Res 37(4):508–513

Brambilla A, Buffoli M, Capolongo S (2019) Measuring hospital qualities. A preliminary investigation on health impact assessment possibilities for evaluating complex buildings. Acta Biomed 90(9S):54–63

Brambilla A, Capolongo S (2019) Healthy and sustainable hospital evaluation-a review of POE tools for hospital assessment in an evidence-based design framework. Buildings. 9(4):76

Colorni A, Tsoukiàs A (2018) What is a decision problem? Designing alternatives. In: Preference disaggregation in multiple criteria decision analysis. Springer, Cham, pp 1–15

Dell'Ovo M, Frej EA, Oppio A, Capolongo S, Morais DC, de Almeida AT (2017) Multicriteria decision making for healthcare facilities location with visualization based on FITradeoff method. In: International conference on decision support system technology. Springer, Cham, pp 32–44. https://doi.org/10.1007/978-3-319-57487-5_3

Dell'Ovo M, Capolongo S, Oppio A (2018a) Combining spatial analysis with MCDA for the siting of healthcare facilities. Land Use Policy 76:634–644

Dell'Ovo M, Frej EA, Oppio A, Capolongo S, Morais DC, de Almeida AT (2018b) FITradeoff method for the location of healthcare facilities based on multiple stakeholders' preferences. In: International conference on group decision and negotiation. Springer, Cham, pp 97–112

Dente B (2014) Understanding policy decisions. In: Understanding policy decisions. Springer, Cham, pp 1–127

Faroldi E, Fabi V, Vettori MP, Gola M, Brambilla A, Capolongo S (2019) Health tourism and thermal heritage. Assessing Italian Spas with innovative multidisciplinary tools. Tour Anal 24(3):405–419

Ferretti V, Montibeller G (2016) Key challenges and meta-choices in designing and applying multicriteria spatial decision support systems. Decis Support Syst 84:41–52

Ferretti V, Pomarico S (2013) Ecological land suitability analysis through spatial indicators: an application of the analytic network process technique and ordered weighted average approach. Ecol Ind 34:507–519

Keeney RL (1992) Value focused thinking. Harvard University Press, Cambridge

Malczewski J, Rinner C (2015) Multicriteria decision analysis in geographic information science. Springer, New York, pp 220–228

Oppio A, Buffoli M, Dell'Ovo M, Capolongo S (2016) Addressing decisions about new hospitals' siting: a multidimensional evaluation approach. Ann dell'Istituto superiore di sanità 52(1):78–87

Rebecchi A, Buffoli M, Dettori M, Appolloni L, Azara A, Castiglia P, D'Alessandro D, Capolongo S (2019) Walkable environments and healthy urban moves: Urban context features assessment framework experienced in Milan. Sustainability 11(10):2778

Riabacke M, Danielson M, Ekenberg L (2012) State-of-the-art prescriptive criteria weight elicitation. Adv Decis Sci

Roy B (1985) Multicriteria methodology for decision analysis. Kluwer Academic Publishers

Roy B (2005) Paradigms and challenges. In Multiple criteria decision analysis: state of the art surveys. Springer, New York, pp 3–24

Chapter 5
Policy Implications. How to Support Decision-Makers in Setting and Solving Complex Problems

Abstract Stakeholders participation in the field of the public decision stimulates learning processes able to generate common knowledge based on shared information. In fact, by including different stakeholders in the decision process different knowledge domains can be integrated. To facilitate this processes, Decision Support Systems (DSSs) have been framed to support stakeholders in decision making for specific purposes. The contribution aims at reflecting on stakeholder participation and to propose a possible participatory process in the context of the location of healthcare facilities based on the methodological framework developed by Simon extended to the scale of Arnstein. Connections of the study within the line of research concerning the "Policy Analytics" perspective are proposed highlighting the importance of the combination of data-driven with value-driven approaches. Moreover, this conclusive chapter will synthetize main achievement and findings of the book.

5.1 Introduction

Complex problems are characterised by a high level of uncertainty (Capolongo et al. 2019). According to the definition of complexity given by Bennet and Bennet (2008), decision problems are complex if they cannot be understood in simple analytic or logical ways, since they may contain several interconnected sub-problems with conflictual interests (Curşeu and Schruijer 2020). In these kind of choices, it could be difficult to define a unique solution since different viewpoints have to be considered and therefore different responses could be generated (Curşeu and Schruijer 2020; Bérard et al. 2017; Bennet and Bennet 2008).

Stakeholder participation has been identified as the key to deal with wicked and ill-structured problems since it enables to open the decision making as well as the evaluation process to different expertise and multiple preferences (Coleman et al. 2017) especially when public decisions have to be taken.

By including different stakeholders in the decision process, analysts can expand their disciplinary boundaries by integrating different knowledge domains, whereas Decision-Makers (DMs) obtain a widely consensus according to a broad set of values, legitimate their choices and define appropriate and participated strategies within the public policy context (Crow et al. 2019). On the analyst side, different forms of collaboration can be identified which promote the interaction of different disciplines in order to solve complex problems and engage several knowledges and the most frequent are: multidisciplinarity; interdisciplinarity and transdisciplinarity. According to Collin (2009), multidisciplinarity refers to different disciplines working on the same topic but independently within their own boundaries; interdisciplinarity is the integration and interaction among different disciplines on the same topic; while transdisciplinarity transcends different disciplines by providing a global synthesis of concepts and methods which can be applied on many topics (Lattuca 2003). In this context, the instance of how to involve different experts into decision-making processes and the assumption that integration between disciplines is an essential feature of the processes of building and expanding knowledge, has emerged.

On the DMs side, to open decision-making processes to several stakeholders stimulates learning processes able to generate common knowledge based on shared information (Berni and Oppio 2015). Within this context, Decision Support Systems (DSSs) have been framed to support stakeholders in decision making for specific purposes, being conceived as a general guide that can be adapted to the special instances of several as well as different contexts (Sarkkinen et al. 2019; Karmperis et al. 2013).

The aim of the contribution is to propose a reflection on stakeholder participation in order to propose a possible participatory process in the context of the location of healthcare facilities (Sect. 5.2) which can bring to the concept of policy analytics (Sect. 5.3) which is further discussed in the conclusions (Sect. 5.4).

5.2   Stakeholder Participation. Involving Stakeholders in Policy Making

By facing complex systems such as the location of public services as healthcare facilities, the involvement of different categories of stakeholder is strategic in order to build a general consensus.

Stakeholder participation addresses the problem of legitimacy in policy making by replacing the more traditional top-down models into a collaborative ones based on the elicitation of values and expectations shared by the actors involved (Berni and Oppio 2015). Participation in public policies has emerged by the demand of transparency and accountability from governments (OECD 2005), being this issue within real world public policy decisions highly political oriented (O'Fairchrallaingh 2010).

Considering the whole policy making process, stakeholders' participation should be placed by DMs not only into the final phase of validation, but it should aid the entire cycle from the setting stage, as a resource able to create innovation and to suggest solutions to face complex problems (O'Fairchrallaingh 2010). By stressing the learning attitude stimulated by participatory processes, stakeholders can work together to

Fig. 5.1 The Arnstein's scale

frame effective and shared strategies and potential opportunities (O'Fairchrallaingh 2010; Chavez and Bernal 2008; Diduck and Mitchell 2003; Fitzpatrick 2006; Van den Howe 2006; Webler et al. 1995).

Talking about the participations and the methodology to stimulate the stakeholders' engagement, the scale of Arnstein (1969) can be considered as a huge support to describe interactions and networks among actors involved in a decision process. As it is possible to appreciate from Fig. 5.1, six main levels are present in the hierarchy and the greater the extent of participation and control over decisions, the smaller the number of possible stakeholder representatives to be engaged in the process. It is possible to detect a base of huge interaction where many different stakeholders have the possibility to be involved, while, by increasing the level of participation, when awareness and experience about the decision problem are required, the number of stakeholder decreases.

Figure 5.1 is not contradictory with what has been previously promoted about the participation, even if it seems that only one actor is in charge to decide, in fact, at the base there should be a large consensus supported by the provision of information otherwise the pyramid collapses. In fact, it can be conceived as the scale of participatory democracy considering the point of view of the attitude of public administrations where citizens are certainly part of the framework and determine the participatory policies of the institutions generally in any local context. Moreover, the scale describes the interaction developed among different levels (Arnstein 1969):

- Provision of information: it legitimates the participation by informing citizens about their responsibilities, possible options, etc. but usually it can be considered as a one-way flow of information;

- Public hearing: it concerns the consultation of citizens' opinion usually carried on through questionnaires and surveys;
- Consultation: it consists in a higher degree of influence but underlies specific knowledges of actors and the priority stated according to the decision problem and context;
- Collaboration: it consists in a partnership between citizens and powerholders. The feasibility and robustness of the partnership is strictly related to the organization of the community and the availability of resources;
- Delegation: it consists in a negotiation between citizens and public officials. It can result in specific decision problems guided by the authority of citizens;
- Self-management: it concerns to control over specific topic but being able to negotiate with external parties.

In line with the Arnstein's Scale, the International Association of Public Participation designed a Public Participation Spectrum (IAP2 2007) organised into five levels, namely inform, consult, involve, collaborate and empower, and structured considering the stakeholder involvement in decision making. Starting from the first level, which involves a merely provision of information, to the last one, where the outcome is strongly influenced by the stakeholder participation, the intensity of interaction increase. The framework developed suggests different forms of influence to the final decision which can be selected considering the project to perform and the stakeholders to engage and within the same projects different levels of interaction could be applied. Given these premises, according to the nature of the policy process an effective participation could suggest a collaborative governance (Nguyen Long et al. 2019), which overcome contestation and conflict between stakeholders to engage in consensus-oriented decision making (Ansell and Gash 2008).

In the decision problem concerning the location of healthcare facilities, according to the Scale of Arnstein, the Public Participation Spectrum and the stakeholder analysis performed in Chap. 3, Fig. 5.2 tries to combine the methodological framework based on Simon extended model with the scale of Arnstein. Compared to the four phases of Intelligence, Design, Choice, and Implementation, another phase, fundamental in public decision, has been added among the last two steps: Decision.

Describing the diagram, it is possible to underline how not all the categories of stakeholders participate to the whole process. Starting from the first phase of Intelligence, where the objectives to achieve are stated, the five categories of stakeholder are taken into consideration (Dente 2014), while in the other phases, when specific competences are required, the number of categories involved decrease until the decision, where only political actor are engaged. In fact, in the design phase, where the problem is framed actors with general interest are not involved while in the choice one, where potential options or strategies are compared they are heard again.

However, it is important to underline how, in order to obtain a general consensus about the final decision to take, political actors need the support and to consult all the other stakeholders, otherwise the process and the pyramid collapses. The last phase that concerns a review of the work will include also experts, actors with

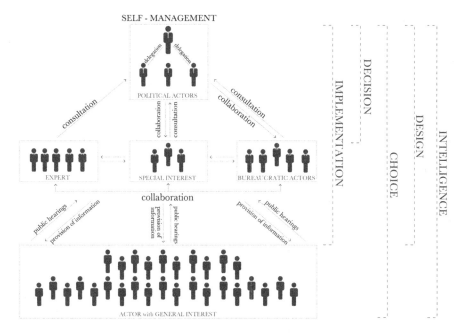

Fig. 5.2 How stakeholders participate to the decision process

special interest and bureaucratic actors since it is necessary to validate the method and different competences are required.

Considering the relations and resources could be exchanged among them, the interactions previously explained have been combined and mixed in order to better describe the decision problem concerning the location of healthcare facilities.

The opinion of actors with special interest is going to be heard from the other categories and then information and decisions should be provided to them. Experts, actors with special interest and bureaucratic actors work on the same level and they should collaborate together in order to achieve the same objective, and their expertise is going to be consulted by political actors to take a decision. Even if they are on different levels, they are strongly interconnected. Moreover, since inside the political category a hierarchy is present, their relation is guided by the delegation.

This brief overview is important in order to understand how a multi-methodological approach proposed for the location of hospitals could really be applied in real life and how could involve all the level of the society to satisfy stakeholders' needs and expectations.

5.3   The Value of Policy Analytics

Policy decisions impact multiple categories of stakeholders and different dimensions of citizen life, i.e. social, economic and cultural. In fact, in policy cycle, conceived as the whole set of actions which lead from the problem setting phases to the solving ones (Daniell et al. 2016), many different actors look for legitimation to participate since directly or indirectly influenced by the decisions. When a real instance of evaluation exists, the policy cycles can be structured into stages to support DMs with the aim of promoting transparency and participation, as in the case of the evaluation framework developed for the location of healthcare facilities. The stakeholders' involvement facilitates learning processes, through the exchange of knowledges, values and resources. This learning model is at the base of the "Policy Analytics" perspective (Tsoukias et al. 2013). With the term Policy Analytics are denoted all the set of methodologies and techniques aimed at aiding stakeholders in any stage of a policy cycle by constructive approaches addressed to support policy decision making and driven by value analysis.

The location of healthcare facilities, the methodology developed to set and solve the problem and the framework structured to promote the participation of different stakeholders can be recognised as an operational contribution to the policy analytics line of research, a vibrant field which lead to enhance the quality of public policy decisions. This approach overcome the criticalities detected from the practice oriented Evidence-Based Policy Making (EBPM) (Blair 1994), based mainly on data and the idea to use evidence to legitimate public policies. In fact, the main characteristics of EMPM concern the availability of high-quality data coherent with the topic under investigation, the collaboration of professionals with different knowledges and the presence of political stakeholders to incentive evidence-based analysis and support decision making processes (Tsoukias et al. 2013; Head 2013). On the contrary, Policy Analytics integrates and combines data-driven with value-driven approaches aimed at considering both data information from the context and objectives stated by stakeholders. Within this context, the legitimation of public policies is not obtained only by the use of evidence but they are constructed together with the categories of stakeholders engaged (Azzopardi Muscat et al. 2020).

5.4   Conclusions

The engagement of stakeholders within policy cycle can improve the effectiveness of public policies.

Complex problems are characterised by a multiplicity of stakeholders with different levels of interest and power, and with their own objectives elicited in relation to their connection with the policy process (Helbig et al. 2015; Capolongo et al. 2016).

Taking into account possible conflicts or coalitions among different groups of stakeholders is a strategic aspect in order to set and face public decision problems by increasing the level of legitimation, transparency and efficiency of decision making processes and governance (Bryson 2004).

Considering the literature analysed on this topic it has been clearly pointed out how the identification of stakeholders to involve is a fundamental phase that further affect the objectives to be achieved which is tailored to the decision context. Despite the uniqueness of each process, some guidelines in relation to the engagement of stakeholders and the different levels of participation can be drawn. The suggested framework can be considered as a first attempt to answer to the necessity of transparency, legitimacy and participation within public decisions.

The contributions proposed within this book try to fill the gap between theory and practice in the field of public decisions with a special attention to healthcare facilities location problems. Given the importance and the complexity of the topic, and its influence on public health, it has been recognised the urgency to bridge this gap by combining different fields of research and multidisciplinary domains. Potential solutions and methodological frameworks have been developed in order to answer to the main open questions of the problem under investigation. First of all, to provide DMs with feasible alternatives, operational recommendations about the research questions have been identified. This gap has been bridged through the development of a DSS (see Chap. 4), able to guide DMs across all the phases of decision processes. Since decision contexts are different, as the territorial and the social ones, DMs need a tool able to be shaped according to specific needs and requirements and to be modified and tailored case by case with the possibility to review all the phases according to emerging issues and instances. Moreover, it supports the engagement of groups of stakeholders by eliciting and including their opinion within the evaluation process.

This book provides a starting point for better understanding how different approaches and methodologies can be applied to support DMs and stakeholders' participation toward better policy choices concerning the location of healthcare facilities. The proposed DSS sheds a light on the relevance of supporting public policies by expanding the boundaries of the decision domain, not providing a finite set of alternative options but generating new and unexpected alternatives by meeting the social and territorial context characteristics.

References

Ansell C, Gash A (2008) Collaborative governance in theory and practice. J Public Adm Res Theor 18(4):543–571

Arnstein SR (1969) A ladder of citizen participation. J Am Inst Plann 35(4):216–224

Azzopardi-Muscat N, Brambilla A, Caracci F, Capolongo S (2020) Synergies in design and health. The role of architects and urban health planners in tackling key contemporary public health challenges. Acta Biomed 91(3-S):9–20

Bennet A, Bennet D (2008) The decision-making process for complex situations in a complex environment, First chapter in Burstein F, Holsapple CW (eds) Handbook on decision support systems

Bérard C, Cloutier LM, Cassivi L (2017) The effects of using system dynamics-based decision support models: testing policy-makers' boundaries in a complex situation. J Decis Syst 26(1):45–63

Berni M, Oppio A (2015) L'Analisi Multicriteri a supporto di procedure di progettazione e pianificazione participate. In: Fattinanzi E, Mondini G (eds) L'analisi Multicriteri Tra Valutazione E Decisione. DEI, pp 27–30

Blair T (1994) Labour party manifesto

Bryson JM (2004) What to do when stakeholders matter: stakeholder identification and analysis techniques. Public Manage Rev 6(1):21–53

Capolongo S, Cocina GG, Gola M, Peretti G, Pollo R (2019) Horizontality and verticality in architectures for health. Technè 17:152–160

Capolongo S, Lemaire N, Oppio A, Buffoli M, Roue Le Gall A (2016) Action planning for healthy cities: the role of multi-criteria analysis, developed in Italy and France, for assessing health performances in land-use plans and urban development projects. Epidemiol Prev 40(3–4):257–264

Chávez BV, Bernal AS (2008) Planning hydroelectric power plants with the public: a case of organizational and social learning in Mexico. Impact Assess Project Appraisal 26(3):163–176

Coleman S, Hurley S, Koliba C, Zia A (2017) Crowdsourced Delphis: designing solutions to complex environmental problems with broad stakeholder participation. Glob Environ Change 45:111–123

Collin A (2009) Multidisciplinary, interdisciplinary, and transdisciplinary collaboration: implications for vocational psychology. Int J Educ Vocat Guidance 9(2):101–110

Crow DA, Albright EA, Koebele EA (2019) Stakeholder participation and strategy in rulemaking: a comparative analysis. State Polit Policy Q 19(2):208–235

Curşeu PL, Schruijer SG (2020) Participation and goal achievement of multiparty collaborative systems dealing with complex problems: a natural experiment. Sustainability 12(3):987

Daniell KA, Morton A, Insua DR (2016) Policy analysis and policy analytics. Ann Oper Res 236(1):1–13

Dente B (2014) Understanding policy decisions. In: Understanding policy decisions. Springer, Cham, pp 1–127

Diduck A, Mitchell B (2003) Learning, public involvement and environmental assessment: a Canadian case study. J Environ Assess Policy Manage 5(03):339–364

Fitzpatrick P (2006) In it together: organizational learning through participation in environmental assessment. J Environ Assess Policy Manage 8(02):157–182

Head BW (2013) Evidence-based policymaking–speaking truth to power? Aust J Public Adm 72(4):397–403

Helbig N, Dawes S, Dzhusupova Z, Klievink B, Mkude CG (2015) Stakeholder engagement in policy development: observations and lessons from international experience. In: Policy practice and digital science. Springer, Cham, pp 177–204

International Association for Public Participation (IAP2) (2007) IAP2 spectrum of public participation. International Association for Public Participation, Thornton

Karmperis AC, Aravossis K, Tatsiopoulos IP, Sotirchos A (2013) Decision support models for solid waste management: review and game-theoretic approaches. Waste Manag 33(5):1290–1301

Lattuca LR (2003) Creating interdisciplinarity: Grounded definitions from college and university faculty. Hist Intellect Cult 3(1):1–20

Nguyen Long LA, Foster M, Arnold G (2019) The impact of stakeholder engagement on local policy decision making. Policy Sci 52(4):549–571

O'Faircheallaigh C (2010) Public participation and environmental impact assessment: purposes, implications, and lessons for public policy making. Environ Impact Assess Rev 30(1):19–27

OECD Organisation for Economic Co-operation and Development (2005) Evaluating public participation in policy making

Sarkkinen M, Kujala K, Gehör S (2019) Decision support framework for solid waste management based on sustainability criteria: a case study of tailings pond cover systems. J Clean Prod 236:117583

Tsoukias A, Montibeller G, Lucertini G, Belton V (2013) Policy analytics: an agenda for research and practice. EURO J Decis Processes 1(1–2):115–134

Van Den Hove S (2006) Between consensus and compromise: acknowledging the negotiation dimension in participatory approaches. Land Use Policy 23(1):10–17

Webler T, Kastenholz H, Renn O (1995) Public participation in impact assessment: a social learning perspective. Environ Impact Assess Rev 15(5):443–463

Printed in the United States
By Bookmasters